The emotional economy of holidaymaking

The emotional economy of holidaymaking

Health, pleasure, and class in Britain, 1870–1918

YAARA BENGER ALALUF

OXFORD
UNIVERSITY PRESS

OXFORD
UNIVERSITY PRESS

Great Clarendon Street, Oxford, OX2 6DP,
United Kingdom

Oxford University Press is a department of the University of Oxford.
It furthers the University's objective of excellence in research, scholarship,
and education by publishing worldwide. Oxford is a registered trade mark of
Oxford University Press in the UK and in certain other countries

First Edition published in 2021

Impression: 1

Published in the United States of America by Oxford University Press
198 Madison Avenue, New York, NY 10016, United States of America

British Library Cataloguing in Publication Data
Data available

Library of Congress Control Number: 2020946919

ISBN 978–0–19–886615–2

DOI: 10.1093/oso/9780198866152.001.0001

Printed and bound in the UK by
TJ Books Limited

Acknowledgements

This book grew out of my doctoral dissertation, which was written at The Center for the History of Emotions at the Max Planck Institute for Human Development and submitted to the Department of History and Cultural Studies at the Free University of Berlin in 2017. I am therefore most indebted to the devoted guidance and generous support of my advisor Ute Frevert, and the insightful comments of the additional members of my doctoral committee: Paul Nolte, Alexander Nützenadel, Sebastian Conrad, and Minu Haschemi Yekani. I am grateful to the financial support of the International Max Planck Research School 'Moral Economies of Modern Societies', without which this research would not have been possible.

I owe a debt of gratitude to Eva Illouz, who had a major influence on my intellectual development and who continues to be a source of inspiration and encouragement. My thanks also go to Margrit Pernau and Stefanie Schüler-Springorum for their support and feedback, to Jan E. Goldstein for the warm hospitality at the Department of History of the University of Chicago, and to Jane M. Adams and Malcolm Neesam for sharing their knowledge and insights.

I thank the series editors, Ute Frevert and Thomas Dixon, for their interest in this project and for the opportunity to publish it as part of the *Emotions in History* series. I am also grateful to the OUP editorial team, and especially to Stephanie Ireland and Sophie Robinson, who helped to see this book through the publication process, to Adam Bresnahan for his superb copy-editing, to Sumintra Gaur, Martin Noble and Esther Benger for their work on the final edits and proofs, and to Rosalba Putrino (Onderosa Indexing) for preparing the index. I would like to thank the anonymous reviewers for their positive endorsement of the manuscript and their constructive suggestions, which helped to shape the book.

I would like to express my sincere gratitude to the archives and libraries whose public records were crucial for my research. I would particularly like to thank Tom Carter at the Blackpool History Centre, James Goddard at the Trades Union Congress Library Collections, Andrew Hill and Samantha Findlay at the Harrogate Library, Angela Kale at the Scarborough Library, Annie Pinder at the Parliamentary Archives, Hasso Spode at the Historical Archive on Tourism in Berlin, and Tom Richardson and Kimberley Starkie at the North Yorkshire County Record Office. Thanks are also due to the staff of the library of the Max Planck Institute for Human Development and the British Library.

I am grateful to Rukmini Barua, Fanny H. Brotons, Edgar Cabanas, Anna Danilina, Daniel Gilon, Çiçek Ilengiz, Amir Minsky, Joseph Ben Prestel, Thomas Rohringer, Daphne Rozenblatt, Marie Schubenz, Mattan Shachak, and Ilaria Scaglia for reading and commenting on different versions or parts of the text, for sharing their thoughts, for providing me with invaluable advice and critical feedback, and for their kindness. I am fortunate to have friends, colleagues, and intellectual partners like you.

Finally, and most of all, I would like to thank my family and my partner for their endless support, encouragement, and love.

Contents

List of figures

List of abbreviations

BL	The British Library
	Add. MSS. Additional Manuscripts
	GRC General Reference Collection
BLP	Blackpool History Centre
BNA	The British Newspaper Archive
GCR	Great Central Railway
GER	Great Eastern Railway
GNR	Great Northern Railway
GWR	Great Western Railway
HAT	Historical Archive on Tourism, The Technical University of Berlin
HRG	Harrogate Library, Local Studies Collection
NYCRO	The North Yorkshire County Record Office
	BPC Book of Press Cuttings
PP	UK Parliamentary Papers
	HC Deb House of Commons debates
	HL Deb House of Lords debates
PQHN	ProQuest Historical Newspapers
SCB	Scarborough Library, Local Studies Collection
TUC	Trades Union Congress Library Collections at London Metropolitan University

Introduction

On 29 September 1911, *The Yorkshire Evening Post* published a letter written by an anonymous 'London Physician'. Returning from a visit to the elite Harrogate Spa, the Londoner complained about both the cure and the attractions provided to the visitors. 'There seems to be a general opinion that the cure is good but might be much better, and the "attractions" of Harrogate... are nil', he claimed. Interestingly, the doctor focused on the amusements provided 'outside the actual baths buildings', during the time that 'the visitors are not making their cures'. Besides the amusements, his detailed letter pointed to a variety of problems at the resort: the cleanliness of the towels, the wages of the workers, and even the disturbing fact that the jackets of the members of the orchestra 'were seldom buttoned up'.[1] This text could have been disregarded as a grumbling letter from a random doctor who returned from a boring holiday if it was not for two points: first, the 'London Physician' is representative of many other physicians of his time who emphasized the significance of amusement at holiday resorts, and second, his letter generated a local storm in a teacup.

'Considerable attention has been aroused at Harrogate by the criticisms of "A London Physician"', reported *The Yorkshire Evening Post*.[2] Among those who responded to the letter was none other than Sir Archibald Boyd-Carpenter, the mayor of Harrogate, who replied to the anonymous doctor in an interview titled 'No Joy Wheels Wanted'. After stressing that he 'disputes entirely that the towels are badly laundried', the mayor asserted that Harrogate does

> not propose to cater for the public on the lines of Blackpool or Morecambe. Patients who come here are supposed to take as much fresh air and rest as possible. That is a part of the treatment. And can it be said that the joy-wheel form of attraction would tend to the patients' repose?[3]

Another commentator from Harrogate echoed the mayor's argument in a letter to the editor, claiming that the 'London Physician' seemed to be 'obsessed with the present craze for excitement in the name of holiday', and 'loses sight of

[1] Anon., 'Wake up, Harrogate! A Visiting Doctor's Plain Criticisms', *The Yorkshire Evening Post*, 29 September 1911, 6/NYCRO DC/HRG/BPC 1911–1912.

[2] Anon., 'No Joy Wheels Wanted. Mayor of Harrogate's Reply to Medical Critic', *The Yorkshire Evening Post*, 30 September 1911, 5/NYCRO DC/HRG/BPC 1911–1912.

[3] Ibid.

The emotional economy of holidaymaking: Health, pleasure, and class in Britain, 1870–1918. Yaara Benger Alaluf, Oxford University Press (2021). © Yaara Benger Alaluf. DOI: 10.1093/oso/9780198866152.003.0001

the important fact that the rest and quietude' enjoyed at Harrogate 'are vital to the cure'.[4]

Why were doctors and politicians debating over the right features of a holiday resort? Why was a physician demanding more attractions while the mayor was defending 'Harrogate's position as a "cure-town"'?[5] And why did the latter refer to the popular working-class seaside resorts Blackpool and Morecambe even though the critic from London never mentioned them? Nor did he ask for a joy-wheel as the mayor seems to imply. This strange case should be understood as a battle in the larger war over the meaning and practices of holidaymaking in late-nineteenth and early twentieth-century Britain. It reflects the debates of the time on the purpose of the holiday (cure or joy) and its methods (treatments or attractions). It hints at assumptions about class and desire, as the visitors of Harrogate apparently wished for rest and quietude, in contrast to the professional Londoner, who advocated for excitement and jollity, allegedly speaking to the tastes of the working-class crowd. It also points to the contested expertise and authority of physicians on the topic of leisure activities. The present book addresses all of these issues by examining the shifting epistemologies of holidaymaking in British resorts between the 1870s and the First World War.

From the mid-nineteenth century onwards, workers' leisure time came to attract considerable interest from politicians, moralists, physicians, and labour movements in Britain. The growing support for workers' holidays was first institutionalized on a national level in The Bank Holiday Act of 1871, the first law to guarantee vacation to adult male workers, and reached its climax in the Holidays with Pay Act of 1938 that granted workers with the right to an annual holiday subsidized by the employer. Gradually, increasing numbers of workers joined the middle and upper classes at local holiday resorts, first for one day trips and later for longer stays, boosting the British holiday industry and marking the dawn of mass tourism. At the same time, the main motivation for visiting watering places—both spas and seaside resorts—shifted from an original focus on physical cure to a search for amusement, fun, and happiness. The story of this profound transformation in British holiday culture is one of the popularization of a commercial leisure sphere based on the conception that all people have the right to have a good time. As will become clear, this new way of thinking was closely related to an increasing scientific interest in emotions and its impact on notions of health, work, and class.

[4] V. C. Mallan, 'Is Harrogate Enterprising? Visitor Thinks Council Should "Let Well Alone". The Rest Cure and Merry-Go-Rounds', *The Yorkshire Evening Post*, 12 October 1911/NYCRO DC/HRG/BPC 1911–1912.

[5] Anon., 'No Joy Wheels Wanted. Mayor of Harrogate's Reply to Medical Critic', *The Yorkshire Evening Post*, 30 September 1911, 5/NYCRO DC/HRG/BPC 1911–1912.

The conditions for the popularization of holidaymaking

Great Britain is generally considered to be the birthplace of modern tourism. The 'Grand Tour', a journey to canonical sites of European culture that sowed the seeds for modern tourism, was established in eighteenth-century Britain for the education of young aristocrats. But Britain had a central role in developing another subtype of tourism: holidaymaking, and more specifically, working-class holidaymaking.[6] Sociologist Erik Cohen defines the holidaymaker as a tourist who '"enjoys" his trip, because it restores his physical and mental powers and endows him with a general sense of well-being'. As we shall see, a long and complex history stands behind this succinct definition.[7] However, from the perspective of the history of tourism, popular holidaymaking did not derive from the Grand Tour, but from a different aristocratic practice: 'taking the waters', meaning bathing in and drinking mineral waters for health purposes.[8] Beginning in the seventeenth century, more and more spas opened up both in Britain as well as on the continent.[9] A key factor in this was the 'scientific revolution' and the increasing trust in the therapeutic qualities of mineral waters, which was reinforced

[6] See, for instance, Eric Zuelow, *A History of Modern Tourism* (London: Palgrave Macmillan, 2015); Susan Barton, *Working-Class Organisations and Popular Tourism, 1840–1970* (Manchester: Manchester University Press, 2005); Fred Inglis, *The Delicious History of the Holiday* (London: Routledge, 2005); John Urry, *The Tourist Gaze* (London: Sage Publications, 2002); Hartmut Berghoff, Barbara Korte, Ralf Schneider, and Christopher Harvie (eds), *The Making of Modern Tourism: The Cultural History of the British Experience, 1600–2000* (London: Palgrave Macmillan, 2002).

[7] Erik Cohen, *Contemporary Tourism: Diversity and Change* (Amsterdam: Elsevier, 2004), 70. For a detailed review of the different definitions of tourism see Urry, *The Tourist Gaze*; Ning Wang, *Tourism and Modernity: A Sociological Analysis* (Oxford: Pergamon, 2000); John Urry, *Consuming Places* (London: Routledge, 1995). In this book I deliberately refrain from defining 'holidaymaking' and 'emotion' from the outset. As conceptual historian Margrit Pernau notes following Reinhart Koselleck's approach: 'Even at one point in time, within one language, the meaning of concepts can never be defined unequivocally: concepts, according to Koselleck, are words whose meaning is contested by social actors.' Margrit Pernau, 'Provincializing Concepts: The Language of Transnational History', *Comparative Studies of South Asia, Africa and the Middle East* 36, no. 3 (2016): 483–99, here 485. Especially due to the fact that 'holiday' is both the central analytical concept *and* the term used in the sources, instead of trying to fix it deductively I consider its fluid meaning empirically in light it use by contemporaneous actors. The term 'emotion' is rarely used in the source material as such, but also as an analytical concept I assert that it should be historicized rather than essentialized by a definition. See Boddice Rob, 'The History of Emotions: Past, Present, Future', *Revista de Estudios Sociales* 62 (2017): 10–5, here 11.

[8] Allan Brodie, 'Towns of "Health and Mirth": The First Seaside Resorts, 1730–1769', in Peter Borsay and John K. Walton (eds), *Resorts and Ports: European Seaside Towns since 1700* (Bristol: Channel View Publications, 2011); Inglis, *The Delicious History of the Holiday*, 38–40; Urry, *The Tourist Gaze*, 4, 17; John K. Walton, *The English Seaside Resort: A Social History 1750–1914* (Leicester: Leicester University Press, 1983), 9.

[9] Jane M. Adams, *Healing with Water: English Spas and the Water Cure, 1840–1960* (Manchester: Manchester University Press, 2015); Peter Borsay, 'A Room with a View: Visualising the Seaside, c. 1750–1914', *Transactions of the Royal Historical Society* 23 (2013): 175–201, here 180; John K. Walton, 'Health, Sociability, Politics and Culture. Spas in History, Spas and History: An Overview', *Journal of Tourism History* 4, no. 1 (2012): 1–14; Phyllis M. Hembry, *The English Spa, 1560–1815: A Social History* (London: The Athlone Press, 1990); Phyllis M. Hembry, *British Spas from 1815 to the Present: A Social History* (New Jersey: Fairleigh Dickinson University Press, 1997).

during the eighteenth century by the scientific study of hydrotherapy.[10] From small springs with very basic amenities, hundreds of watering places were converted into health resorts. The most famous of these resorts were established in Bath, Harrogate, and Scarborough and became popular holiday destinations for the highest strata of society.[11] As spas were also considered to be social centres, efforts were made to not only cure, but also to entertain visitors, and tennis courts, libraries, theatres, and ballrooms gradually grew into important parts of social life at the spas.[12] However, as will be elaborated in Chapter 3, up until the second half of the nineteenth century, the recreation facilities at the health resorts were quite basic and unorganized, and, more importantly, they were not seen as the main reason for visiting the resort, but as secondary to the therapeutic objectives.

Although bathing and swimming in the ocean, in lakes and in rivers existed as forms of recreation prior to the eighteenth century, a network of commercial holiday resorts in watering places was established and gained in popularity only after physicians started advocating cold water bathing and uses of salt-water as a cure for sickness. Thus, during the eighteenth and the first half of the nineteenth century, the seaside resort, just like the spa, was considered to be a place of healing rather than of pleasure.[13] In Britain, the new 'rush to the sea' led to the rapid growth of seaside towns and new seaside resorts, which in many ways followed the model of the spas. Although seaside resorts were regarded at first as a version of the inland spas, they actually posed a major commercial threat to many spas and marked a significant step towards the entirely new phenomenon of mass holidaymaking.[14]

[10] Adams, *Healing with Water*; George Weisz, 'Spas, Mineral Waters, and Hydrological Science in Twentieth-Century France', *Isis* 92, no. 3 (2001): 451–83; Hembry, *British Spas from 1815 to the Present*; John K. Walton, 'The History of British Spa Resorts: An Exceptional Case in Europe?', *Transportes, Servicios y Telecomunicaciones* 20 (2011): 138–57, here 148.

[11] It should be noted that, curiously, the British spa was had significantly less influence—socially, culturally and economically than it had in other European countries. See, for instance Walton, 'The History of British Spa Resorts'; Adams, *Healing with Water*, 211–2.

[12] Karl E. Wood, *Health and Hazard: Spa Culture and the Social History of Medicine in the Nineteenth Century* (Newcastle: Cambridge Scholars Publishing, 2013), 1–2; Walton, 'Spas in History, Spas and History'; Hembry, *British Spas from 1815 to the Present*; Brodie, 'The First Seaside Resorts, 1730–1769', 23–4; John Alfred Ralph Pimlott, *The Englishman's Holiday: A Social History* (Hassocks: Harvester Press, 1976 [1947]), 23–4, 29–30.

[13] Walton, 'The History of British Spa Resorts', 140–2, 144; Brodie, 'The First Seaside Resorts, 1730–1769', 18–20, 27; Alain Corbin, *The Lure of the Sea: The Discovery of the Seaside in the Western World, 1750–1840* (Berkeley: University of California Press, 1994); Harold J. Perkin, 'The "Social Tone" of Victorian Seaside Resorts in the North-West', *Northern History* 11, no. 1 (1976): 180–94, here 183–4; Orvar Löfgren, *On Holiday: A History of Vacationing* (Berkeley: University of California Press, 1999), 116–8; on the similar process in Scotland, see Alastair Durie, 'Medicine, Health and Economic Development: Promoting Spa and Seaside Resorts in Scotland c. 1750–1830', *Medical History* 47, no. 2 (2003): 195–216, here 197–8.

[14] John Alfred Ralph, Pimlott, 'The English Seaside', *English* 10, no. 59 (1955), 167. See also Brodie, 'The First Seaside Resorts, 1730–1769'; Urry, *The Tourist Gaze*, 17; Hembry, *British Spas from 1815 to the Present*, 6. On the development of British resorts during the seventeenth and eighteenth centuries as a particular urban form focused on the supply of health and leisure see Peter Borsay, 'From Port to

Holiday resorts were founded throughout Britain, inland and along the coast, mainly in the second half of the eighteenth century. However, the Lancashire coast was the first place where workers and members of the lower classes visited resorts, followed by Yorkshire, Staffordshire, and Birmingham.[15] Slowly, a new hierarchy of 'social tone' developed between the different sites. By the 1870s, some resorts, such as Blackpool, were clearly labelled by the upper classes as inferior and vulgar.[16] The coast was visited by both men and women, even though it was mostly only men who went swimming. Up until the early twentieth century bodies were seldom exposed outside of the water, and mixed bathing was considered 'immoral' and forbidden. The Victorian bathing machines helped to protect bathers' privacy, until men and women had the freedom to bathe together.[17] Perhaps this change was related to holidaymaking's radical shift towards becoming a mass phenomenon, with more than half of the English population visiting the seaside at least once a year in the beginning of the twentieth century.[18]

But how did this happen? Although the present book does not ask what enabled the popularization of resorts, but focuses on the shifting epistemologies of holidaymaking, it is nonetheless crucial to review the historical causes that allowed for this phenomenon: What enabled and motivated more and more Britons, including growing numbers of members of the working classes, to visit holiday resorts?[19] John Alfred Ralph Pimlott was the first historian to address these questions and was also the first to provide a thorough account of secular holiday's evolution in nineteenth and early twentieth century Britain.[20] John K. Walton followed Pimlott's pioneering research on British holidaymaking, focusing primarily on the seaside holiday industry.[21] Fortunately, the last decade witnessed a renewed interest also in the history of the British spa with the works of Peter Borsay, Jane Adams, Walton, and others.[22] These studies have pointed to a number of

Resort: Tenby and Narratives of Transition, 1760–1914', in Peter Borsay and John K. Walton (eds), *Resorts and Ports: European Seaside Towns since 1700* (Bristol: Channel View Publications, 2011).

[15] John K. Walton, *The British Seaside: Holidays and Resorts in the Twentieth Century* (Manchester: Manchester University Press, 2000), 51–2; Perkin, 'The "Social Tone" of Victorian Seaside Resorts in the North-West', 183–4.

[16] Perkin, 'The "Social Tone" of Victorian Seaside Resorts in the North-West'; Urry, *The Tourist Gaze*, 16.

[17] Walton, *Holidays and Resorts*, 40; Borsay, 'Visualising the Seaside, c. 1750–1914', 197–8.

[18] Urry, *The Tourist Gaze*, 18.

[19] It should be clarified from the outset that the incorporation of the working classes into the resort holiday culture was far from homogeneous and varied particularly from one region to another.

[20] Pimlott, *The Englishman's Holiday*; Pimlott, 'The English Seaside'.

[21] Walton, *The English Seaside Resort*; Walton, *Holidays and Resorts*.

[22] Phyllis Hembry's books *The English Spa 1560–1815: A Social History* and *British Spas from 1815 to the Present: A Social History* were until recently the most detailed accounts of the British spa. However, due to her early death, the latter is far from being satisfactory. See also John K. Walton, 'Review of *British Spas from 1815 to the Present: A Social History* by Phyllis Hembry', *Urban History* 26, no. 2 (1999): 299–300. Recent histories of British spa resort and water cure include: Adams, *Healing with Water*; Peter Borsay, 'Town or Country? British Spas and the Urban–Rural Interface', *Journal of*

conditions for the popularization of holidaymaking and the growth of the British holiday resort in the late nineteenth century: free time, supply, purchasing power, transportation, and demand. Even more than other forms of leisure, holidaymaking depends first and foremost on time available for travel and for the stay at the destination. Time is thus the basic condition.[23] Indeed, early closing regulations, half-holidays, regularized days of leisure and in some professions even week-long holidays, were achieved in Britain, and in particular northern England, from the mid-nineteenth century.[24] Comparable working hours and holiday legislation were implemented in France and Germany only in the first decades of the twentieth century.[25] As will be explored in detail in Chapter 2, holidays for workers were first secured in the first half of the nineteenth century through collective bargaining with employers. This was translated into legislation during the second half of the century. The Bank Holiday Act of 1871 provided a half-holiday on Saturdays and a secular holiday on the first Monday of August, and represents the first step in a long legislative process that led to the reduction of the work week and facilitated wider working populations to make holidays away from home.[26] At first, these holidays were unpaid. Only by the late nineteenth century did paid holidays spread to professional and white-collar jobs, and only by the First World War did some industrial employers start granting them to their workers.[27]

However, and especially from a bourgeoisie perspective, reducing working hours was not enough, since many of the limited recreational opportunities for workers in the fast-growing industrial cities were considered immoral or non-productive (e.g. alcohol consumption, games, and races). Hence, from the mid-nineteenth century onwards, politicians and moralists undertook efforts to 'civilize' working-class recreational styles. Hence, besides the creation of leisure time, politicians and civil society actors sought to guide workers to consume a specific

Tourism History 4, no. 2 (2012): 155–69; Walton, 'Spas in History, Spas and History'; Walton, 'The History of British Spa Resorts'. See also Wood, *Spa Culture and the Social History of Medicine in the Nineteenth Century;* Ronan Foley, Healing Waters: Therapeutic Landscapes in Historic and Contemporary Ireland (Farnham: Ashgate Publishing, 2010); Alastair J. Durie, *Water is Best: The Hydros and Health Tourism in Scotland, 1840–1940* (Edinburgh: John Donald, 2006).

[23] Peter Borsay, *A History of Leisure: The British Experience since 1500* (Hampshire: Palgrave Macmillan, 2006), 192–3; John K. Walton, 'The Demand for Working-Class Seaside Holidays in Victorian England', *The Economic History Review* 34, no. 2 (1981): 249–65.

[24] Walton, 'The Demand for Working-Class Seaside Holidays in Victorian England'; Urry, *The Tourist Gaze*, 19–20.

[25] Peter N. Stearns, *Lives of Labor: Work in a Maturing Industrial Society* (New York: Holmes and Meier Publishers, 1975).

[26] As will be explained in Chapter 2, this was particularly significant for Southern workers, since workers in the north of England had succeeded in securing the traditional wakes holidays during the first half of the century.

[27] Borsay, *A History of Leisure*, 193; Walton, *Holidays and Resorts*, 59.

kind of *supply* that was considered 'rational' and 'civilized'.[28] These forms of recreation were based on established bourgeois habits such as concerts, lectures, sports, gardening, and excursions to the seaside.[29] Even without diving into theories about the relation between leisure and consumption,[30] we can agree that another necessary condition for the rise of mass holidaymaking was *purchasing power*, meaning that larger sectors of society had to have high enough wages and sufficient savings to spend on travel and recreation.[31] Indeed, the British national income per person quadrupled over the run of the nineteenth century, and by the 1870s most working-class households could also save money or rely on savings clubs for their leisure activities. With the rise of living standards, larger publics could engage in leisure activities, which in turn drew more visitors to the holiday resorts.[32]

A fourth condition was *access* to the desired holiday destination.[33] Whereas travel in the first half of the nineteenth century was still slow, uncomfortable, and expensive, the railway made transportation much faster and more affordable. Gladstone's Railway Regulation Act of 1844, which contracted the railway companies to supply transportation to low-income passengers, initiated the democratization of travel and ensured the accessibility of destinations previously considered remote.[34] Consequently, the most accessible resorts, such as Blackpool, Southend, and Morecambe were the most popular. This trend was

[28] See, for instance Helen Meller, *Leisure and the Changing City 1870–1914* (Oxon: Routledge, 2013), 11; William J. Baker, 'The Leisure Revolution in Victorian England: A Review of Recent Literature', *Journal of Sport History* 6, no. 3 (1979): 76–87, 81–2; Brad Beaven, *Leisure, Citizenship and Working-Class Men in Britain, 1850–1945* (Manchester: Manchester University Press, 2005); Borsay, *A History of Leisure*, 87–90.

[29] Trent S. Newmeyer, '"Moral Renovation and Intellectual Exaltation": Thomas Cook's Tourism as Practical Education', *Journal of Tourism and Cultural Change* 6, no. 1 (2008): 1–16; Peter Bailey, *Leisure and Class in Victorian England: Rational Recreation and the Contest for Control, 1830–1885* (Buffalo: Routledge & Kegan Paul Books, 1978).

[30] On this issue see, for instance Peter Borsay, 'Economy', in *A History of Leisure: The British Experience since 1500* (Hampshire: Palgrave Macmillan, 2006); Jon Stobart, Andrew Hann, and Victoria Morgan, *Spaces of Consumption: Leisure and Shopping in the English Zown, c. 1680–1830* (Oxon: Routledge, 2013); Douglas J. Goodman and Mirelle Cohen, 'From Consumption to Consumer Culture', in Douglas J. Goodman and Mirelle Cohen (eds), *Consumer Culture: A Reference Handbook* (Santa Barbara: ABC-CLIO, 2004); Peter N. Stearns, 'The Explosion of Consumerism in Western Europe and the United States', in *Consumerism in World History: The Global Transformation of Desire* (London: Routledge, 2006); Rosalind H. Williams, 'The Dandies and Elitist Consumption', in *Dream Worlds: Mass Consumption in Late Nineteenth-Century France* (Berkeley: University of California Press, 1991); Thorstein Veblen, *The Theory of the Leisure Class* (Oxford: Oxford University Press, 2007 [1899]).

[31] Borsay, *A History of Leisure*, 24; Walton, 'The Demand for Working-Class Seaside Holidays in Victorian England'.

[32] On saving clubs and 'going-off' clubs, especially for financing holidays on the Wakes Week see Robert Poole, 'Oldham Wakes', in John K. Walton and James Walvin (eds), *Leisure in Britain 1780–1939* (Manchester: Manchester University Press, 1983), 71–98; Barton, *Working-Class Organisations and Popular Tourism*, 76–8.

[33] Walton, 'The Demand for Working-Class Seaside Holidays in Victorian England'.

[34] Urry, *The Tourist Gaze*, 21; Gary S. Cross and John K. Walton. *The Playful Crowd: Pleasure Places in the Twentieth Century* (New York: Columbia University Press, 2005), 14–6.

also supported by Thomas Cook's package tours, which deployed the new passenger railways to attract thousands of visitors to the seaside resorts each year.[35] By the 1880s, all seaside resorts had direct railway access.[36] The fact that inland spas were significantly less accessible by train was probably one of the reasons for the higher popularity of the coast, and perhaps also for the relative decline of the British spa industry as a whole.[37]

Indeed, mass transportation was an important condition for the expansion of the holiday resort industry. However, as Harold Perkin shows, chronologically it is evident that it was not a necessary condition, and it still does not explain the motivation for holidaymaking.[38] Pimlott was therefore right to question the direct causal relation between the two, noting that 'it does not automatically follow from the existence of a modern system of transport that it will be used for pleasure, still less for holidays away from home'.[39] This is especially true keeping in mind that even with Gladstone's Railway Act, railway travel was a considerable expense for the average member of the working classes.[40] The same should be said about the relation between free time and capital: the means do not determine the particular desires. In other words, an additional necessary condition for the increasing popularity of holidays in watering places was *demand*. This is to say that a significant number of potential holidaymakers had to have a desire to spend their free time and income at holiday locales in the first place.[41] Interestingly, in the case of working-class holidaymaking this desire was also quite unique to Britain, where by the mid-nineteenth century workers counted affordable leisure among their demands for better working conditions. In contrast, up until the first decade of the twentieth century, workers on the continent focused on conditions at the workplace or on material needs, such as family grocery shopping.[42]

As Susan Barton shows, workers were not passive consumers of a product supplied to them by the upper classes or by manufacturers. Worker movements and unions had an active role not only in controlling working hours and

[35] Claudia Bell and John Lyall, *The Accelerated Sublime: Landscape, Tourism and Identity* (Westport: Praeger Publishers, 2002), 13.

[36] Walton, *Holidays and Resorts*, 73.

[37] Walton, 'The History of British Spa Resorts', 144–7.

[38] Perkin, 'The "Social Tone" of Victorian Seaside Resorts in the North-West', 184–5; see also Borsay, *A History of Leisure*, 24, 29; Pimlott, *The Englishman's Holiday*, 78; Douglas A. Reid, 'The "Iron Roads" and "the Happiness of the Working Classes" The Early Development and Social Significance of the Railway Excursion', *The Journal of Transport History* 17, no. 1 (1996): 57–73.

[39] Pimlott, *The Englishman's Holiday*, 78.

[40] The actual accessibility of railway travel to the working classes and the regional differences in the nominal price of excursion fares are still under debate. See for instance Timothy Leunig, 'Time Is Money: A Re-assessment of the Passenger Social Savings from Victorian British Railways', *The Journal of Economic History* 66, no. 3 (2006): 635–73; Jack Simmons, *The Victorian Railway* (London: Thames and Hudson, 1991), 74–5, 318–23; Walton, 'The Demand for Working-Class Seaside Holidays in Victorian England'.

[41] Borsay, *A History of Leisure*, 89, 179; Walton, 'The Demand for Working-Class Seaside Holidays in Victorian England'.

[42] Stearns, *Lives of Labor*, 252.

conditions, but in converting holidaymaking into a major leisure practice.[43] Still, it remains unclear why they turned to this specific form of leisure. The common explanation assumes that lower classes tend to embrace the tastes and habits of the upper classes once they become affordable.[44] This kind of emulation can also be interpreted as an active claim to rights. From this point of view, as wages rose, the working classes came to conceive of the types of holidaymaking that were already established for the upper strata as a social right rather than as a luxury.[45] However, not all scholars accept the 'class emulation' theory. For example, Hartmut Berghoff argues that British holidaymaking resulted out of a combination of influences coming both from above, such as romantic views of travel as an expression of personal freedom, and from below, such as folk festivals.[46] In the same vein, John K. Walton claims that while the 'class emulation' theory works for the democratization of spas, the seaside holiday must be read as an independent contribution of the British working classes.[47] It is safe to say that because an established network of spa and seaside resorts already existed, these places were the most obvious and available destinations for those who wished to get away from the city.[48]

From physical cure to emotion management

The five conditions outlined above explain how the popularization of holiday-making was made possible, they do not explain, though, why as more and more people chose to visit local holiday resorts, the original purposes and characteristics of these resorts themselves changed. This change is frequently described as a process through which the former medicinal uses of watering places—both spas and seaside resorts—gave way to the new objective of fun and amusement. Although 'previously the emphasis had been placed on health',[49] as the turn of the century approached, the medical motivation faded and holidays were

[43] Barton, *Working-Class Organisations and Popular Tourism.*

[44] Borsay, *A History of Leisure*, 25; Colin Campbell, 'Conspicuous Confusion? A Critique of Veblen's Theory of Conspicuous Consumption', *Sociological Theory* 13 (1995): 37–47; Pimlott, *The Englishman's Holiday*, 74.

[45] Susan Barton, 'Why We're All Going on a Summer Holiday: The Role of the Working-Class Organisations in the Development of Popular Tourism, 1850–1950', PhD diss., De Montfort University, 1999, 61; Beaven, *Leisure, Citizenship and Working-Class Men in Britain*, 44.

[46] Hartmut Berghoff, 'From Privilege to Commodity? Modern Tourism and the Rise of the Consumer Society', in Hartmut Berghoff et al. (eds), *The Making of Modern Tourism: The Cultural History of the British Experience, 1600–2000* (London: Palgrave Macmillan, 2002).

[47] John K. Walton, 'British Tourism between Industrialization and Globalization – An Overview', in Berghoff et al. (eds), *The Making of Modern Tourism.*

[48] Walton, 'The Demand for Working-Class Seaside Holidays in Victorian England', 249; Urry, *The Tourist Gaze*, 18.

[49] Darren Webb, 'Bakhtin at the Seaside: Utopia, Modernity and the Carnivalesque', *Theory, Culture & Society* 22, no. 3 (2005): 121–38, here 128.

increasingly identified 'with pleasure rather than health'.[50] Walton, who dedicated his career to the history of the British seaside, points to 'a remarkable transition from a preoccupation with health, self-improvement and decorum to an eager pursuit of pleasure and amusement', which took place as 'idleness on holiday became more a matter for enjoyment than guilt'.[51] Regarding the spa resort he similarly argues that its 'concentration on the medical gaze became diluted over time'.[52] Orvar Löfgren, in his book on the history of vacationing, also claims that the Victorian seaside resort reflects a 'gradual transformation from health to fun'. This transformation, he argues, led people to visit the seaside 'not as a prescribed cure but as a secret pleasure... a refuge in which to enjoy a true life—simple and natural'.[53]

These are only a few examples of the common description of the development of popular resort holidaymaking as 'a shift from therapy to leisure, from health to pleasure'.[54] It is this narrative—which I will refer to from now on as the 'narrative of depathologization'—that I seek to challenge in this book. My argument proceeds by placing emotions at the centre of the analysis. A simplified overview of the historical transformation at stake might help make this new approach more comprehensible. We can see that the alleged process of depathologization occurred hand in hand with two shifts in the holiday industry: the clientele was quickly broadening, no longer confined to aristocrats and disabled persons, and resorts were offering more amusements and less medicinal treatments. Historians of holidaymaking therefore need to clarify the correlation between the democratization of holidaymaking, the change in practices or methods of holidaymaking, and the rise of a new purpose for holidaymaking as an 'unalloyed pursuit of pleasure'.[55]

The modification of methods applied at watering places has been studied by historians of medicine David Cantor and George Weisz, who point to a fading confidence in the curative powers of water, especially following the First World War. This, they claim, occurred both because of the growing scientific criticism of hydrotherapy and because of developments in medical science that provided more

[50] Walton, *The English Seaside Resort*, 183. See also Walton, 'Spas in History, Spas and History', 9.

[51] Walton, *The English Seaside Resort*, 182–3.

[52] Walton, 'Spas in History, Spas and History', 9.

[53] Löfgren, *On Holiday: A History of Vacationing*, 118. Löfgren's book 'travels widely in time and space' (p. 9); however, his study of seaside holidaymaking is based mostly on examples from Sweden and the British Victorian seaside resort.

[54] Annick Cossic, 'Introduction', in Patrick Galliou and Annick Cossic (eds), *Spas in Britain and in France in the Eighteenth and Nineteenth Centuries* (Newcastle: Cambridge Scholars Press, 2006), xix.

[55] Cross and Walton, *The Playful Crowd*, 5. Indeed, first signs of both shifts can be traced back to the eighteenth century, but certainly not in the magnitude they were manifested in the second half of the nineteenth century and later. This will be further explained throughout the book, and particularly in Chapter 3.

reliable and effective therapies.[56] This narrative, which was reiterated also in reference to earlier periods by Pimlott and Walton, has been recently put into question, most famously by Jane Adams.[57] Even if we assume that this process started earlier, only reaching its peak in the interwar period, this account fails to explain the correlation between the popularization and democratization of holiday resorts and the changes in their methods. We might speculate that the working classes had less faith in the water cure, but we still need to understand how a new objective emerged and why 'fun' turned out to be this new objective.

A different thesis, represented most prominently by Walton, focuses on the new clientele and emphasizes workers' need to escape from industrial urban life.[58] Walton reads holidays in watering places as offering a release from the monotonous work and harsh discipline of the factories, the crowded urban space, and 'from boredom and frustration'.[59] This approach suffers from a different shortcoming, as it fails to clarify exactly why visiting places that had been regarded as therapeutic for almost two centuries was considered the right solution for these problems,[60] considering that other forms of resisting or coping with the new conditions of urban industrialization were possible.[61] We also need to understand why resorts visited chiefly by the upper classes, such as Harrogate, also gradually marketed their supply in terms of happiness and amusement while maintaining their exclusivity.[62]

In sum, while some accounts do not take into serious consideration the characteristics of the new clientele (lower-middle and working classes), others overlook the original goal of the destination (cure). In contrast, the argument of this book combines both points and goes as follows: the perception of watering places as sites of cure did not disappear; rather, it was the pathology they were supposed to treat that changed. Once the emotional implications of modernity and work came to be viewed as potentially pathological, the management of emotions was turned into a therapeutic practice, and people came to see positive emotional experiences as healing. At a time when recreation was controlled along class lines, the seemingly neutral and scientific framing of emotions enabled idle

[56] George Weisz, 'Historical Reflections on Medical Travel', *Anthropology & Medicine* 18, no. 1 (2011): 137–44; Weisz, 'Spas, Mineral Waters, and Hydrological Science in Twentieth-Century France', 479–80. David Cantor, 'The Contradictions of Specialization: Rheumatism and the Decline of the Spa in Inter-War Britain', *Medical History: Supplement* 34, no. 10 (1990): 127–44.

[57] Pimlott, *The Englishman's Holiday,* 106; Walton, 'The History of British Spa Resorts', 148; Adams, *Healing with Water.*

[58] Walton, *Holidays and Resorts,* 27; Urry, *The Tourist Gaze,* 18–9; John K. Walton and James Walvin, 'Introduction', in John K. Walton and James Walvin (eds), *Leisure in Britain 1780–1939* (Manchester: Manchester University Press, 1983).

[59] Cross and Walton, *The Playful Crowd,* 63.

[60] The fact that sea bathing was practiced by the lower classes also in earlier centuries not necessarily in relation to health does not explain, as we shall see, the therapeutic branding of holidays in watering places for the working classes in particular.

[61] See for instance Stearns, *Lives of Labor,* 241–7, 252.

[62] This will be further elaborated in Chapter 4.

holidays to be legitimized for all classes. We cannot properly understand the phenomenon of holidaymaking by treating health and pleasure as opposites or by searching for a transition from health to pleasure. Rather, we must examine the changes in the *meaning* of each of these concepts. As we shall see, a more accurate description would be not that health was replaced by leisure, but that emotion management was increasingly replacing the physical cure. These dynamics, in turn, informed a new epistemology of holidaymaking.

As discussed above, previous studies on the sharp changes undergone by local British holidaymaking tend to focus above all on material factors, including increasing wages and technological innovation.[63] But the history of the popularization of holidaymaking went hand in hand with changes in its *emotional appeal and purpose*. This point has not received serious theoretical attention, historians of British holidaymaking generally only treating it in very general formulas without further elaboration, such as a 'change of mental attitude'[64] or a 'revolution in experience, thinking and perception' that led to the evolution of a 'widespread acceptance of the view that going on holiday was good for one'.[65] But if workers were looking for an 'excuse for having some fun',[66] and the seaside in particular became 'a new kind of emotional space',[67] then why have emotions hardly been regarded as a central variable in research on the popularization and democratization of holidaymaking?[68]

For fun and happiness to become the chief objectives of holidaymaking, a new demand for these emotions had to appear and receive social justification. For this reason, it is the contention of this book that the rise of popular holidaymaking in

[63] See also Walton, 'The Demand for Working-Class Seaside Holidays in Victorian England'; Peter Borsay and John K. Walton, 'Introduction: The Resort–Port Relationship', in Peter Borsay and John K. Walton (eds), *Resorts and Ports: European Seaside Towns since 1700* (Bristol: Channel View Publications, 2011), 1.

[64] Pimlott, *The Englishman's Holiday*, 23.

[65] Urry, *The Tourist Gaze*, 26.

[66] John Beckerson and John K. Walton, 'Selling Air: Marketing the Intangible at British Resorts', in John K. Walton (ed.), *Histories of Tourism: Representation, Identity and Conflict* (Clevedon: Cromwell Press, 2005), here 66.

[67] Löfgren, *On Holiday: A History of Vacationing*, 118.

[68] Corbin's book *The Lure of the Sea* (1994) made the case for a change in the emotional appraisal of watering places between the mid-eighteenth and mid-nineteenth centuries. In doing so it became one of the earliest works relating nature, and sea bathing in particular, to shifts in perceptions and meanings of emotional and bodily experiences, or what would now be called 'the history of emotions'. The book was ambitious and path breaking, and yet had various shortcomings, both in its approach to the study of emotions in history and in its analysis of local histories, as John K. Walton convincingly argued in his review of Corbin's book (John K. Walton, 'Review of *The Lure of the Sea: The Discovery of the Seaside in the Western World 1750–1840*, by Alain Corbin', *European History Quarterly* 25, no. 3 (1995): 450–2). Corbin's study is rich and stimulating, combining a variety of sources: from religious and mythical texts to medical literature and from travellers' diaries to the visual arts and architecture, and his comprehensive account for emotional expressions is remarkable. However, his analysis is confined to the discourses of the upper classes about the sea, regarded mainly as landscape and environment. The relation between these discourses and other social strata (were they aware of it? did they contest it? are there parallel shifts in the discourses of lower classes? or did the elite discourse 'trickle down'?) as well as their influence on more tangible aspects of the sea and its shore (ports, resorts, commercial amusements) remain obscure.

late nineteenth-century British resorts depended to a large extent on shifts in the way people understood emotions and the emotional self. Whereas understandings of emotions and their impact on social institutions as well as on notions of identity are receiving more and more historiographical attention,[69] the impact of these processes on tourism cultures has not yet been systematically examined.[70] In his seminal book *The Tourist Gaze*, sociologist John Urry notes that '[m]aking theoretical sense of "fun, pleasure and entertainment" has proved a difficult task for social scientists'.[71] This book seeks to meet the challenge by combining the history of emotions and the sociology of commodification to better grasp the crystallization of a new epistemology of holidaymaking in British Victorian and Edwardian resorts.

This shift in perspective opens up new questions: What were the emotional justifications underlying the democratization of holidaymaking? Which moral and emotional assumptions regarding human nature and needs were at the basis of this process? What influence did scientific studies exert on the political debates on work and leisure and on practices of holidaymaking? Why was it that as the number of working-class visitors increased, the explicit objective of holiday resorts changed from providing physical cure to supplying fun and amusement? Was this an essential affinity between social class and emotional demand or were the resorts reflecting a particular perception of the 'improper classes'? Were there gender-specific differences in how emotions were conceived of, and was there a difference in demand and experience between men and women at holiday resorts? How are emotional experiences converted into commodities and how do people learn how to consume them?

[69] See, for instance Joanna Bourke, *The Story of Pain: From Prayer to Painkillers* (Oxford: Oxford University Press, 2014); Rob Boddice, *The Science of Sympathy: Morality, Evolution, and Victorian Civilization* (Urbana: University of Illinois Press, 2016); Sabine Donauer, 'Emotions at Work – Working on Emotions: The Production of Economic Selves in Twentieth-Century Germany', PhD diss., Free University of Berlin, 2013; Susan Matt, 'Current Emotion Research in History: Or, Doing History from the Inside Out', *Emotion Review* 3, no. 1 (2011): 117–24; Ute Frevert, *Emotions in History: Lost and Found* (Budapest: Central European University Press, 2011).

[70] It should be noted that emotional and affective aspects of spas, both in Britain and in the continent were researched, mainly in the work of Jill Steward, but not in terms of an epistemic change. See Jill Steward, 'The Affective Life of the Spa', in David Picard and Mike Robinson (eds), *Emotion in Motion: Tourism, Affect and Transformation* (Farnham: Ashgate Publishing, 2012); Jill R. Steward, 'Moral Economies and Commercial Imperatives: Food, Diets and Spas in Central Europe: 1800–1914', *Journal of Tourism History* 4, no. 2 (2012): 181–203.

[71] Urry, *The Tourist Gaze*, 7. Almost two decades later, Urry did intend to tackle the place of emotions in geography and tourism more directly, although not as part of a thorough historical argument. I will discuss this work in the next section. See John Urry, 'The Place of Emotions within Place', in Joyce Davidson, Liz Bondi, and Mick Smith (eds), *Emotional Geographies* (Aldershot: Ashgate, 2007).

Doing history of emotions and consumption

The history of Victorian and Edwardian[72] holidaymaking accounts for changes in perceptions of emotional life and practices just as much as it tells the story of the rise of a particular commercial sphere in which emotional experiences are consumed. By analysing the emotional economy of holidaymaking, the present book addresses two issues that have been often overlooked within two vibrant fields of research: emotions in the theory of tourism, and consumption in the history of emotions.

The last decades have seen a substantial increase in interest in emotions as an object of historical inquiry, both as a category for historical analysis and as a variable that affects history.[73] One of the central approaches for the history of emotions follows the path that Carol and Peter Stearns paved in their study of emotional standards. Their method distinguishes between the normative aspects of emotions on the one hand and subjective emotions on the other. The Stearns conceived of the structural dimension of emotions as 'emotionology', claiming that emotional standards vary according to historical and social context.[74] At first glance, analysing emotional norms might seem to fit the aims of this book, since it focuses on the ways in which implicit cultural regulations are created and how they shape knowledge about and experiences of emotions. However, this approach has its weak points. For instance, to whom does a particular normative framework apply to in a given period and culture and how are its limits defined? How are emotional norms produced, learned, and embodied? Is the historian able to empirically observe emotions and go beyond the mere analysis of standards of emotional expression?[75]

By focusing on emotion knowledge: its production, circulation and contestation, and not exclusively on 'norms' and 'institutions', I seek to challenge the sharp distinction drawn in the emotionology approach between individual

[72] The period covered in this book includes three eras in British history: the Victorian, the Edwardian, and the reign of George V until the end of the First World War. It is therefore hard to stick to a periodization represented by monarch names. As a rule, when using the term 'Victorian' I refer to the period until 1901 and when referring to 'Victorian and Edwardian', I include also George's reign and the First World War.

[73] On the research of the history of emotions see, for instance Jan Plamper, *The History of Emotions: An Introduction* (Oxford: Oxford University Press, 2015); Susan J. Matt and Peter N. Stearns (eds), *Doing Emotion History* (Urbana: University of Illinois Press, 2013).

[74] Carol Z. Stearns and Peter N. Stearns, 'Introduction', in *Anger: The Struggle for Emotional Control in America's History* (Chicago: University of Chicago Press, 1989); Peter N. Stearns and Carol Z. Stearns, 'Emotionology: Clarifying the History of Emotions and Emotional Standards', *The American Historical Review* 90, no. 4 (1985): 813–36.

[75] For further theoretical elaboration see Benno Gammerl, 'Transitory Feelings? On Challenges and Trends within the History of Emotions', *Contemporanea* 17, no. 2 (2014): 335–44; Margrit Pernau, 'Male Anger and Female Malice: Emotions in Indo-Muslim Advice Literature', *History Compass* 10, no. 2 (2012): 119–28; Fay Bound Alberti, 'Introduction: Medical History and Emotion Theory', in Fay Bound Alberti (ed.), *Medicine, Emotion and Disease, 1700–1950* (New York: Springer, 2006); Joanna Bourke, *Fear: A Cultural History* (London: Virago Press, 2005), 73–6.

experiences and institutional standards. First, my point of departure is that there is no single and uniform 'normative institution' that dictates emotional standards. Thus, in order to explore how emotion knowledge (i.e. meanings and practices of emotions) is created and circulated, I will analyse the relations between contingent and contested systems of knowledge, social institutions, and individuals. This will be done in four different spheres: scientific research, politics and legislation, the resort industry, and holidaymakers' experiences.

Second, in contrast to Stearns and Stearns, my analysis does not favour the normative context over the subjective experience. Although they do call for a historical examination of 'peoples' efforts to mediate between emotional standards and emotional experience',[76] they lack tools for studying the individual level: how actors learn, embody, and express emotions, and how these emotions are interpreted and described. Following more recent approaches to emotional experience as dependent on a *dispositif*[77] or *habitus*,[78] I focus on the construction of the emotional subject by detailing its encounters with the emotion knowledge available while at the same time exposing the role of individuals as agents of emotion knowledge production. This approach also demands pinpointing moments when the norms 'failed'. Addressing failure will allow me to evaluate the degree of correspondence between institutional regulations and individual experience and expose the power relations behind the constitution of emotional standards.[79]

Finally, I am interested in the *value* given to particular emotional states, rather than in the formation of emotional standards or feeling rules. In her book *Emotions in History* Ute Frevert uses the concept 'emotional economy' regarding emotions as a resource 'framed within an economy that depends on, and answers to, cultural expectations and social needs'.[80] This approach allows us to make sense of why the value of certain emotions underwent certain historical changes: it can help facilitate our understanding of why some emotions are attributed with more significance in particular historical periods or societies while others seem to disappear, and why some are seen to be crucial for the individual or the society while others are condemned. Additionally, the concept helps the historian account both for the role that social institutions play in the coordination of this economy as well as for the ways it is learned, embodied, and modified by individual members of society. Hence, it simultaneously acknowledges the individual aspect

[76] Stearns and Stearns, 'Emotionology', 825.

[77] See Eva Illouz, 'Introduction: Emodities or the Making of Emotional Commodities', in Eva Illouz (ed.), *Emotions as Commodities: Capitalism, Consumption and Authenticity* (London: Routledge, 2017).

[78] See Monique Scheer, 'Are Emotions a Kind of Practice (and Is That What Makes Them Have a History)? A Bourdieuian Approach to Understanding Emotion', *History and Theory* 51, no. 2 (2012): 193–220.

[79] On the idea of focusing on failure in the research of the history of emotions, see William M. Reddy, 'Against Constructionism: The Historical Ethnography of Emotions', *Current Anthropology* 38, no. 3 (1997): 327–51; Boddice, *The Science of Sympathy*, 21–2; Scheer, 'Are Emotions a Kind of Practice', 211.

[80] Frevert, *Emotions in History*, 212.

of emotions while highlighting the influence that institutions have both on the social evaluation and standardization of emotions as well as on the ways in which people are educated to approach emotional experience, behaviour, and expression.

Emotional economy, as presented by Frevert, refers to a system in which the *meaning and value* of emotions are negotiated. However, the contemporary polysemy of the term 'economy' allows for a second interpretation of emotional economy as a process through which emotions can be transformed into commodities.[81] This idea—that emotions can be and have been converted into actual objects for consumption—was recently developed by sociologist Eva Illouz.[82] The emotional economy as a site of the commodification of emotions provides a powerful framework for studying the consumption of emotional experiences in the holiday industry and in tourism in general.

The link between economy and emotions has become an object of intense scrutiny in the last decades. Illouz brings together the various interconnections between psychological repertoires and the capitalist market under the umbrella term 'emotional capitalism', which she defines as 'a culture in which emotional and economic discourses and practices mutually shape each other'.[83] The concept of 'emotional capitalism' can help us decipher the multifarious relations between emotions and consumption, including the manipulation of consumers' emotions in order to motivate consumption and the impact of consumption on individual identity. In this sense, analysing emotional capitalism compels us to think about emotions not merely as leading people to consume, but as being constituted through acts of consumption.[84] An additional link between emotions and consumption is the production of emotions as commodities. This is not to say that, with time, commodities we buy acquire emotional meaning, but that some commodities are 'deliberately designed to generate emotions and to be consciously consumed for that reason'.[85] The crucial implication of the attainment of emotional desires through the market is that the emotional sphere is being increasingly framed and organized by economic concepts and commercial actors and vice versa.[86]

[81] The conclusions chapter will elaborate further this point.

[82] Eva Illouz (ed.), *Emotions as Commodities: Capitalism, Consumption and Authenticity* (London: Routledge, 2017).

[83] Eva Illouz, *Cold Intimacies: The Making of Emotional Capitalism* (Cambridge: Polity Press, 2007), 5.

[84] Eva Illouz and Yaara Benger Alaluf, 'Le capitalisme émotionnel', in Alain Corbin, Jean-Jacques Courtine, and Georges Vigarello (eds) *Histoire des émotions vol. 3: De la fi n du XIXe siècle à nos jours (1890–2013)* (Paris: Seuil, 2017). On emotions as a motivational force for consumption from a history of emotions perspective see Susan J. Matt, *Keeping Up with the Joneses: Envy in American Consumer Society, 1890–1930* (Philadelphia: University of Pennsylvania Press, 2003).

[85] Yaara Benger Alaluf and Eva Illouz, 'Emotions in Consumer Studies', in Frederick F. Wherry and Ian Woodward (eds), *The Oxford Handbook of Consumption* (Oxford: Oxford University Press, 2019).

[86] Illouz (ed.), *Emotions as Commodities.*

The scant research on the emotional aspects of holidaymaking is not only an empirical lack—it also has broader theoretical implications.[87] The essence of the tourist experience has been the subject of much controversy.[88] Especially from the 1990s onwards, in addition (and often in relation) to the question of authenticity, many sociologists and historians of tourism emphasize the commercial character of tourism as well as of the holiday experience and address directly the intangible products of this leisure industry. Watson and Kopachevsky, most famously addressed 'tourism as an extension of the commodification of modern social life under capitalism', and argued that the tourist experience 'includes the "consumption" of signs, symbols, cultural experiences'.[89] The semiotic approach to tourism led some later theorizations to conflate 'object' and 'commodity', 'consumption' and 'experience'.[90] Whereas a more sensitive and critical apprehension of the tourist experience developed via the 'performativity turn' in tourism studies since the 2000s, this body of literature is generally oblivious to questions of

[87] Some preliminary thoughts on this matter appeared in Yaara Benger Alaluf, '"It Is All Included—Without the Stress": Exploring the Production of Relaxation in Club Med Seaside Resorts', in Eva Illouz (ed.), *Emotions as Commodities: Capitalism, Consumption and Authenticity* (London: Routledge, 2017).

[88] See Daniel Boorstin, 'From Traveler to Tourist: The Lost Art of Travel', in *The Image: Or What Happened to the American Dream* (New York: Vintage Books, 1992 [1961]); Dean MacCannell, 'Staged Authenticity: Arrangements of Social Space in Tourist Settings', *American Journal of Sociology* 79, no. 3 (1973): 589–603; Louis Turner and John Ash, *The Golden Hordes: International Tourism and the Pleasure Periphery* (London: Constable & Robinson, 1975); Erik Cohen, 'A Phenomenology of Tourist Experiences', *Sociology* 13, no. 2 (1979): 179–201; Donald L. Redfoot, 'Touristic Authenticity, Touristic Angst, and Modern Reality', *Qualitative Sociology* 7, no. 4 (1984): 291–309; Urry, *The Tourist Gaze;* Shuai Quan and Ning Wang, 'Towards a Structural Model of the Tourist Experience: An Illustration from Food Experiences in Tourism', *Tourism Management* 25, no. 3 (2004): 297–305.

[89] Llewellyn G. Watson and Joseph P. Kopachevsky, 'Interpretations of Tourism as Commodity', *Annals of Tourism Research* 21, no. 3 (1994): 643–60, here 649–50; see also John Urry, 'The "Consumption" of Tourism', *Sociology* 24, no. 1 (1990): 23–35; for more general observations regarding the semiotic approach in consumer studies and its contributions and shortcomings for the research of emotions and consumption, see Benger Alaluf and Illouz, 'Emotions in Consumer Studies'; Illouz, 'Introduction: Emodities or the Making of Emotional Commodities'.

[90] For instance, in *The Accelerated Sublime* Bell and Lyall consider the origin of tourism to be in the merging of natural landscape and 'the commercial', thus arguing that the value of tourism is primarily aesthetic and not material. '[A] central commodity in contemporary tourism', they claim, is the experience of the sublime (Bell and Lyall, *The Accelerated Sublime*, 4–5, 24); Anne Friedberg sees travel and mass tourism as a form of 'commodity-experience', Anne Friedberg, *Window Shopping: Cinema and the Postmodern* (Berkeley: University of California Press, 1993), 3, 59. A somewhat more nuanced analysis of the tourist experience, usually based on economic theories, and particularly on the concept of 'experience economy' as developed by Pine and Gilmore, can be found in tourism marketing and management or in consumer behaviour fields. Joseph Pine and James H. Gilmore, *Experience Economy: Work is Theatre & Every Business a Stage* (Boston: Harvard Business School Press, 1999); See, for instance Edgar Cabanas, 'Experiencing Designs and Designing Experiences: Emotions and Theme Parks from a Symbolic Interactionist Perspective', *Journal of Destination Marketing & Management* (in press); Scott A. Cohen, Girish Prayag, and Miguel Moital, 'Consumer Behaviour in Tourism: Concepts, Influences and Opportunities', *Current Issues in Tourism* 17, no. 10 (2014): 872–909; Richard Sharpley and Philip R. Stone, 'Introduction: Experiencing Tourism, Experiencing Happiness?', in Richard Sharpley and Philip R. Stone (eds), *Contemporary Tourist Experience: Concepts and Consequences* (London: Routledge 2012).

commodification and consumption.[91] And yet, as Shuai Quan and Ning Wang observe: 'Despite the growth of literature on the tourist experience, a fundamental issue remains puzzling: What are the components that constitute the tourist experience per se?'[92] To this I add: What exactly is being commodified in tourism, and more specifically, what is that vacationers consume while on holiday?

John Urry's influential concept of the 'tourist gaze' offers one answer, namely, that tourists consume not only services but aesthetic signs, and are in fact engaged in ongoing visual consumption of places and sights.[93] Referring directly to holidaymaking, sociologist Ning Wang suggests a different answer. Arguing that the meaningful experience of the holiday is dependent on the commodification of time, Wang implies that holidaymaking is essentially the consumption of 'free time'.[94] Urry and Wang both emphasize that the tourist experience is, in substance, an experience of difference, thus defining tourism as a cultural response to modern everyday life, and especially regulated and routinized work.[95] The affinity between this 'difference' and emotions is present in both these approaches, but it is either considered as given or explained in general terms as an outcome of 'modernity'.[96] Furthermore, both Wang and Urry note that within the tourist experience emotions and consumption are somehow linked. For Urry, '[p]laces are emotionally pleasurable because they are sites of intense and heightened consumption', but, unfortunately, he does not explain this causal relation.[97] Wang, in contrast, highlights the tension between tourists' motivation and experience, informed by Eros, and the tourism industry, informed by Logos.[98] However, the seemingly puzzling coexistence of emotional authenticity and consumerist practices, subjective pleasures and commercial industries, remains unsolved.

[91] Tim Edensor, 'Performing Tourism, Staging Tourism: (Re)Producing Tourist Space and Practice', *Tourist Studies* 1, no. 1 (2001): 59–81. See also Carl Cater and Paul Cloke, 'Bodies in Action: The Performativity of Adventure Tourism', *Anthropology Today* 23, no. 6 (2007): 13–6; Simon Coleman and Mike Crang (eds), *Tourism: Between Place and Performance* (New York: Berghahn Books, 2002); Stephen Harwood and Dahlia El-Manstrly, 'The Performativity Turn in Tourism', *University of Edinburgh Business School Working Paper Series* 12, no. 5 (2012): 1–18; on emotional performativity in consumer studies, see Benger Alaluf and Illouz, 'Emotions in Consumer Studies'.

[92] Quan and Wang, 'Towards a Structural Model of the Tourist Experience'.

[93] Urry, *The Tourist Gaze*.

[94] Wang, *Tourism and Modernity*, especially Chapter 5.

[95] Wang, *Tourism and Modernity*, 91–2; Urry, *The Tourist Gaze*, 1–2. This theorization resembles Peter Borsay's claim that the 'other', and particularly the different and 'exotic', is one of the elemental features of leisure in general. See Borsay, *A History of Leisure*, 1–8, 176.

[96] Urry observes that changes in emotion knowledge brought by Romanticism and new forms of 'visual desire' were an important precondition for the development of modern tourism and for the creation of the emotional experience of place (Urry, *The Tourist Gaze*, 20; Urry, 'The Place of Emotions within Place', 77–8). Wang, more methodically, frames his discussion of tourism under the modern distinction between a space of 'Logos' and a space of 'Eros', or as a dialectical response to a process of rationalization (Wang, *Tourism and Modernity*, especially Chapter 2). Discussing particularly the experience of different time he mentions that '[i]f working time is informed by a work ethic (doing things), then leisure or holiday time is dictated by the need for relaxation, change, play, and freedom' (Wang, *Tourism and Modernity*, 110).

[97] Urry, 'The Place of Emotions within Place', 77–8.

[98] Wang, *Tourism and Modernity*, 41–2.

I seek to push this discussion further. I am in agreement with the conception of tourism, and holidaymaking in particular, as an experience which is constituted by difference, i.e. in contrast to a certain 'mundane' experience (be it everyday routines, city life, wage labour, etc.). However, I maintain that such an experience of contrast is not only closely linked, but rather contingent upon, historically specific emotional meanings.[99] Accordingly, I argue that the primary object being consumed in holidaymaking is an *emotional experience* that differs from emotional experiences associated with routine and work, which have been culturally classified as negative. The history told in this book is therefore an early example of what Illouz dubs an 'emodity': '*the co-production of emotions and commodities*'.[100] The spatial and temporal aspects perceived by Urry and Wang, respectively, as the core of the tourist experience are explored in the present book with the aim of unravelling the history through which a particular meaning of contrast was generated and popularized.

Combining the history of emotions with the sociology of commodification, my conception of emotional economy extends that of Frevert by introducing the element of consumption and varies from Illouz's emodity in its historical focus and scope. The commodification of emotions presupposes their constitution as objects, which is achieved only once 'emotional experience is organized, labelled, classified, and interpreted'.[101] In other words, the ability to commodify emotional experiences depends on an existing (albeit contested) epistemology of emotion. In this sense, the production of emotional experiences of holidaymaking is not confined to the resort, but is part of a broader history of emotion knowledge. However, there is no reason to assume that these processes are unconnected. I am in agreement with Illouz that consumerist practices and emotion knowledge are mutually produced, a point that will be empirically demonstrated throughout the book. In order to account for this twofold process, I define 'emotional economy' as the circulation of values ascribed to emotions, in the course of which emotions are culturally produced as meaningful objects *and* industrially produced as commodities, thus generating subjects and institutions that are simultaneously emotional and economic.[102]

[99] Benger Alaluf, 'Exploring the Production of Relaxation in Club Med Seaside Resorts', 42–3; Illouz and Benger Alaluf, 'Le capitalisme émotionnel', 71.

[100] Illouz, 'Introduction: Emodities or the Making of Emotional Commodities', 7 (emphasis in original).

[101] Eva Illouz, *Consuming the Romantic Utopia: Love and the Cultural Contradictions of Capitalism* (Berkeley: University of California Press, 1997), 3.

[102] See also Benger Alaluf and Illouz, 'Emotions in Consumer Studies'; Sara Ahmed, 'Affective Economies', *Social Text* 22, no. 2 (2004): 117–39.

Structure of the book

Since my goal is to unravel the interconnections between emotion knowledge and experience, morality and a specific industry, the chapters of this book follow a thematic structure. They explore the emotional aspects of holidaymaking between the last third of the nineteenth century and the early twentieth century by focusing on four topics: the medical community's views of holidaymaking; the justifications underlying holiday legislation; the development of the resort industry; and the experiences of holidaymaking. The depathologization narrative, that is the assumption that modern holidaymaking was the product of a deviation from the traditional therapeutic role of watering places and its substitution with a search of fun and pleasure has never been confirmed by a rigorous analysis of contemporaneous medical literature.[103] Chapter 1 will meet this challenge and show that holidaymaking continued to be perceived by the British medical community as a therapeutic practice at the turn of the century, but that there were significant developments in the conceptualization of the pathology, the patient, and the cure itself. I will show how the emerging emotional pathologization of modernity, and particularly urban work, led to the conception of 'the worker' as a vulnerable emotional subject, irrespective of gender or occupation. The chapter outlines the emotional history of conditions such as neurasthenia and overwork, which played a central role in the development of psychology as an independent discipline and which should be seen as the ancestors of stress and burnout.[104] It then explores how people thought overwork could be cured. The sources analysed, mainly medical texts, reveal that holidaymaking at watering places continued to be viewed as a method of healing throughout the period. I will demonstrate that the theorization of 'change' as a means to manage emotions undergirded the new take on holidaymaking as an efficient way to help people manipulate their emotions and avoid the negative effects of work.

It should be stressed that with 'medical discourse of emotions', I do not necessarily refer to the discipline of psychology. In fact, as will be explained, the Victorian scientific interest in emotions and their relation to health and social conditions preceded the institutionalization of psychology as a distinct discipline, and was carried out in other fields of research, predominantly neurology and psychiatry. In this sense, contemporaneous sources challenge the claim that it was only after the First World War that 'amusement itself...had a "psychological"

[103] In *Healing with Water*, Adams provides an exhaustive account of 'the cure' based on an incredibly rich range of medical sources. However, my interest lies elsewhere. I am not interested in the treatments offered at watering places, but on the wider theories and rationalizations connecting between concepts of healing and of holidaymaking.

[104] See, for instance David Wainwright and Michael Calnan, *Work Stress: The Making of a Modern Epidemic* (Buckingham: Open University Press, 2002), 34; Mark Jackson (ed.), *Stress in Post-War Britain, 1945–85* (Oxon: Routledge, 2015).

justification'.[105] Although psychology as a discipline only began being institutionalized in Britain at the turn of the century, attempts to empower psychological approaches and particularly to identify the relation between modernity, emotions, and health can be traced to the 1870s.

An immense body of literature exists on the transformations that took place in the British sciences and medicine during the eighteenth and nineteenth centuries, and excellent work has been conducted on the flourishing of new discourses about emotions during this period.[106] Especially instructive for the history of emotions are the works of Kurt Danziger and Thomas Dixon, who analysed trends in the understanding of human mental phenomena by tracing lexical changes throughout eighteenth- and nineteenth-century Britain.[107] Particularly relevant for my interests here is the claim that scientific debates of the nineteenth century gave birth to the category of 'emotions' as a secular, non-moral semantic field more universal in scope than previous concepts like passions, appetites, and sentiments.[108] Because it addresses the production of knowledge about emotions in the Victorian and Edwardian medical profession, my research is related to the work of Danziger, Dixon, and others. Nevertheless, a few clarifications are in place. First, unlike Danziger and Dixon, I am not interested in 'psychological thought' in general, which certainly existed inter alia in theological and philosophical contexts; rather, I deliberately focus on the medical approach, although the different contexts sometimes converge.[109] Second, whereas Western psychological thought is founded in various fundamental categories, such as the soul, cognition, drives and motivation, personality, and others,[110] I am interested in one specific category of psychological thought—the emotions. Finally, since my focus is on the medical knowledge that was being generated starting from the second half of the nineteenth century, I will follow the language used in the source

[105] Cantor, 'The Contradictions of Specialization', 132–3.

[106] See for instance Alberti, *Medicine, Emotion and Disease*; Thomas Dixon, *From Passions to Emotions: The Creation of a Secular Psychological Category* (Cambridge: Cambridge University Press, 2003); Rick Rylance, *Victorian Psychology and British Culture, 1850–1880* (Oxford: Oxford University Press, 2000); Kurt Danziger, *Naming the Mind: How Psychology Found Its Language* (London: Sage, 1997); Roy Porter, *Disease, Medicine and Society in England, 1550–1860* (Cambridge: Cambridge University Press, 1995); Andrew Scull (ed.), *Madhouses, Mad-Doctors, and Madmen: The Social History of Psychiatry in the Victorian Era* (Philadelphia: University of Pennsylvania Press, 1981); Walter E. Houghton, *The Victorian Frame of Mind, 1830–1870* (New Haven: Yale University Press, 1957).

[107] Dixon, *From Passions to Emotion*; Danziger, *Naming the Mind*. See also Ute Frevert et al., *Emotional Lexicons: Continuity and Change in the Vocabulary of Feeling 1700–2000* (Oxford: Oxford University Press, 2014).

[108] Thomas Dixon, 'Patients and Passions: Languages of Medicine and Emotion, 1789–1850', in Fay Bound Alberti (ed.), *Medicine, Emotion and Disease, 1700–1950* (New York: Springer, 2006), 25–6; Dixon, *From Passions to Emotion*, 4.

[109] Alberti, 'Introduction: Medical History and Emotion Theory', xxii–xxiii; Rylance, *Victorian Psychology*; Dixon, *From Passions to Emotion*, 9, 24–5.

[110] Danziger, *Naming the Mind*, 5.

material, without engaging in discussions of other concepts such as passion or appetite.

Chapter 2 examines parliamentary papers, documents from workers' unions, moralist writings, and publications in the press in order to enquire into how political actors sought to legitimize workers' holidays. Although the focus of the book is not working time per se, being that any holidaymaking industry on a national scale is dependent upon visitors who have a number of consecutive free days to spend at the resorts, it is crucial to understand the ways that long holidays were legitimized for members of all classes. Urry is correct in claiming that 'by the Second World War... Holidays had become almost a marker of citizenship, a right to pleasure'.[111] But what enabled this process? Many historians have tried to understand what led to the reduction of working time and the concomitant rise of holiday legislation that took place in most Western countries starting from the end of the nineteenth century,[112] and much research has been dedicated to Britain, where this process started, and which became the model for other countries wishing to emulate the 'British week'.[113] A large part of this literature focuses on the influence that macro-economic conditions exerted on the interests of employers and workers by trying to unravel the relationship between time (work, leisure) and money (profit, wages).[114] Significantly less energy has been spent studying factors beyond the worker–employer nexus and the public debate on working hours and workers' holidays.[115] I will argue that health and emotional well-being had a central role in the political debate on workers' free time.

The chapter studies the impact that medical knowledge—with its newly gained hegemonic authority—as well as shifting perceptions of health, gender, and emotions had on holiday legislation, placing an emphasis on the gradual shift in interest from physical to emotional conditions. Similarly, I will discuss how the

[111] Urry, *The Tourist Gaze*, 26.

[112] Gary S. Cross, 'The Quest for Leisure: Reassessing the Eight-Hour Day in France', *Journal of Social History* 18, no. 2 (1984): 195–216; Baker, 'The Leisure Revolution in Victorian England', 78; John Lowerson and John Myerscough, *Time to Spare in Victorian England* (Hassocks: Harvester Press, 1977), 16; Stearns, *Lives of Labor*, 247–8.

[113] Michael Huberman, 'Working Hours of the World Unite? New International Evidence of Worktime, 1870–1913', *The Journal of Economic History* 64, no. 4 (2004): 964–1001; Gary S. Cross, *A Quest for Time: The Reduction of Work in Britain and France, 1840–1940* (Berkeley: University of California Press, 1989); Stearns, *Lives of Labor*; Manfred A. Bienefeld, *Working Hours in British Industry: An Economic History* (London: Weidenfeld and Nicolson [for the] London School of Economics, 1972); E. P. Thompson's seminal article 'Time, Work-Discipline, and Industrial Capitalism', *Past and Present* 38, no. 1 (1967): 56–97 unfortunately does not coincide with the time frame of my research.

[114] Huberman, 'Working Hours of the World Unite'; Gary S. Cross, 'Time, Money, and Labor History's Encounter with Consumer Culture', *International Labor and Working-Class History* 43 (1993): 2–17; Keith McClelland, 'Time to Work, Time to Live: Some Aspects of Work and the Reformation of Class in Britain, 1850–1880', in Patrick Joyce (ed.), *The Historical Meanings of Work* (Cambridge: Cambridge University Press, 1987); Bienefeld, *Working Hours in British Industry*.

[115] Significant exceptions being: Bailey, *Rational Recreation*; Arthur J. McIvor, 'Employers, the Government, and Industrial Fatigue in Britain, 1890–1918', *British Journal of Industrial Medicine* 44, no. 11 (1987): 724–32.

pathologization of the worker helped make legislation more egalitarian and how it ultimately facilitated the inclusion of the working classes into the established middle-class holiday culture. I will also address a question long overlooked in previous research: Why were watering places considered to be the ideal destination for workers' holidaymaking? As will become clear, this question reflects the main difference between previous research on urban 'rational recreation'[116] and the phenomenon of holidaymaking in nature, and points to the impact that the traditional therapeutic role of watering places had on the construction of spa and seaside resorts as central holiday destinations for all classes.

Following up on the previous two chapters' exploration of views on holidaymaking in the medical and political spheres, Chapter 3 turns to the objectives of the resort industry. Starting from the last decades of the nineteenth century, the British holiday resort industry went through rapid growth and commercial development. The chapter will examine this process by drawing on source material that includes internal documents of the local corporations, resort publications (guides and advertisements), vacationers' accounts, and medical literature that refers directly to holiday practices. After explaining the commercialization of the resort and describing the different actors in this economy, I will analyse the dynamics of health, pleasure, and social class, arguing that the emotionally loaded encounter between different social classes at the resort was a core aspect of the shifting practices of holidaymaking. This chapter differs from previous work on the British holiday resort. My aim is not to write an urban or social history of a unique type of town or an economic history of a specific business, nor is it my intention to offer another description of life at the resort. Rather, my goal is to identify the objectives of the British resort industry and their transformations over time in relation to epistemologies of holidaymaking. Hence, I focus on what was seen as the goal of the holiday and the explicit purpose of different holiday activities.

The chapter also introduces new analytical categories to the existing literature. British spa and seaside resorts have received a respectable share of attention in the literature on leisure consumption, being seen variously as a pioneering stepping stone in Western tourism,[117] as a health establishment and site of sociability for the wealthy elites (mostly in the case of the spa),[118] as a unique working-class leisure institution (mostly in the case of the seaside resort),[119] and as a particular case of urban development.[120] As important as this research is, it has by definition

[116] Bailey, *Rational Recreation.*

[117] Inglis, *The Delicious History of the Holiday*; Berghoff et al. (eds), *The Making of Modern Tourism*; Urry, *The Tourist Gaze.*

[118] Walton, 'The History of British Spa Resorts', 138–57; Borsay, 'Town or Country? British Spas and the Urban–Rural Interface', 156–7.

[119] Barton, *Working-Class Organisations and Popular Tourism.*

[120] Walton, *The English Seaside Resort*; Fred Gray, *Designing the Seaside: Architecture, Society and Nature* (London: Reaktion Books, 2006); Borsay, 'Town or Country? British Spas and the Urban–Rural Interface'.

two central blind spots. First, by focusing, too often, on one social class, it ignores analytical categories beyond class, such as emotions. Second, by describing the seaside resort as both a successor of the spa industry and a distinct phenomenon, it has failed to capture how these two institutions co-informed one another.

In order to overcome these challenges, I have elected for a different approach. First, I follow Joanna Bourke's claim that emotionological research cannot a-priori ascribe particular emotions to specific social groups;[121] thus, I do not presuppose that what shaped the objectives for holidaymaking was necessarily the class of the visitors or the type of resort (inland spa or seaside resort). Second, by researching both spas and seaside resorts, I intend to shed new light on the ongoing negotiation of the meaning and methods of holidaymaking. In order to comprehend the changes in holiday practices from 'taking the waters' to commercial amusements, the chapter focuses on three watering places that were among the most popular holiday destinations in Victorian and Edwardian Britain, but which differed in their characteristics: Harrogate, an aristocratic inland spa; Scarborough, a spa and seaside resort with mixed social clientele; and Blackpool, the exemplary working-class seaside town. The chapter concludes by pointing to the impact of the therapeutic rationalization of recreational activities on the resort industry, which, perhaps counterintuitively, gave way to its emotionalization that will be further explored in the next chapter.

Chapter 4 sets forth the investigation of the resort from the previous chapter by taking a closer look at the actual product supplied at resorts. The chapter shows that by the turn of the century, both spas and seaside resorts were explicitly claiming to supply emotional experiences. It also analyses the effects that this change had on notions of gender and class. One of the chapter's key claims is that the emotionalization of holidaymaking did not challenge its therapeutic function; rather, the crucial change was that the therapeutic framing of amusement opened holidaymaking to the lower classes, while at the same time paving the way for physicians to become involved in various aspects of the holiday industry.

Through the analysis of resort publicity, popular literature, and texts written by vacationers, the second part of the chapter explores some of the challenges faced by resorts in their new function as an emotional industry. I will demonstrate that in order to provide emotional change, the resort industry had to adjust itself to all kinds of unstable perceptions of the emotional meaning of modernity, nature, and morality, and their concrete expressions. In this sense, and in contrast to common assumptions about consumerism, I will argue that the kind of consumption practiced in holidaymaking was not entirely subordinated to and manipulated by the production system; rather, the value of the product was dependent on the broader emotional economy.

[121] Bourke, *Fear: A Cultural History*, 353–5.

After examining how a new perception of emotions and of the emotional worker shaped the meaning of the holiday for doctors, legislators, reformers, and resort authorities, Chapter 5 sets to examine how these changing epistemologies were negotiated by resorts' visitors. How and to what extent the therapeutic discourse on holidaymaking was accessible to potential holidaymakers and what were their own interpretations? Drawing on popular stories and poems about holidaymaking, vacationers' accounts, and advice literature the chapter analyses the reciprocal relations between emotion knowledge, economic activity, and subjectivity, and thematizes the experience of holidaymaking and the various roles individuals played in the emotional economy of holidaymaking. As Susan Barton rightly argues, most of the historical research of British holidaymaking is mainly concerned with the ways in which higher classes and the industry manipulated workers' demand on the one hand and with the effect of the democratization of leisure on the tourism industry on the other. Few studies, however, focus on workers' expectations and on how they actively shaped the holiday culture and were shaped by it.[122] Even when focusing on holiday practices and experiences of the middle and upper classes on holiday, no research has been dedicated to the construction of the holidaymaker as an emotional subject. Rejecting the notion that people have an 'essential emotional self',[123] I pursue a performativity approach that hones in on both the process of construction of the emotional subject and its own negotiation of the meaning of emotional experiences.[124] Empirical material for analysing this relation can be found in instances in which individual experiences did not correspond to emotional expectations. Pinpointing moments of disappointment and boredom, I will show that holidaymakers of different genders and classes were aware of the medical view of holidaymaking, even though they did not always identify with them or experience what they promised.

Furthermore, the chapter explores the performative nature of the holiday experience. As mentioned earlier, the argument that tourism is 'a form of performance' is not new, but to the best of my knowledge it has not been affirmed through historical research.[125] While Tim Edensor argues that 'particular tourist contexts generate a shared set of conventions about what should be seen, what should be done and which actions are inappropriate',[126] this chapter uncovers *how*

[122] Barton, *Working-Class Organisations and Popular Tourism.*

[123] Alberti, 'Introduction: Medical History and Emotion Theory', xvi.

[124] Eva Illouz and Yaara Benger, 'Emotions and Consumption', in Dan Cook and J. Michael Ryan (eds), *The Wiley-Blackwell Encyclopedia of Consumption and Consumer Studies* (Oxford: John Wiley & Sons, 2015); Pierre Bourdieu, *Outline of a Theory of Practice* (Cambridge: Cambridge University Press, 1977).

[125] Edensor, 'Performing Tourism, Staging Tourism'. See also Cater and Cloke, 'Bodies in Action: The Performativity of Adventure Tourism'; Coleman and Crang, *Tourism: Between Place and Performance*; Harwood and El-Manstrly 'The Performativity Turn in Tourism'.

[126] Edensor, 'Performing Tourism, Staging Tourism', 60.

tourist contexts generated practices and *how* practices of holidaymaking were translated into emotional experiences that were conceived as personal and authentic, in a particular time and place.[127] This includes exposing the ways in which holidaymakers learned how to feel, how to express their emotions in a commercial public sphere and how to consume emotional experiences. I will show this was done both by taking a disciplinary view of themselves and others during the holiday and by producing and consuming emotion knowledge, that is information about the is-and-ought of holiday experience. I will argue that vacationers did not merely learn how and what to feel while on holiday. As much as the holiday shaped its subjects, their emotions, and their values, the latter had a role in moulding the experience at the resort. In this sense, this chapter continues the discussion opened in the previous chapter on the relation between emotions and consumption and questions the dichotomous neo-Marxist perception of production-consumption relations. It will be explained that in an emotional economy, it is much more difficult to identify a one-way movement or a clear distinction between producer and consumer, as the industry is conditioned by its ability to fulfil consumers' desires, which are in turn related to wider and unstable epistemologies.

The thematic approach and the theoretical drive of this book might raise objection on the part of some historians. I am in agreement with Peter Borsay in that 'it is social and cultural formations and processes that matter' and that 'overemphasis upon *when* change occurred deflects from adequate theorizing about *what* changes happened, and *why*'.[128] It is not the main intention of this book to suggest a chronology of the rise of popular holidaymaking, less so of tourism or commercial leisure in general. The emotionalization of the holiday resort, which is the process explored in this book, occurred in relation to other histories, most of which, as we have seen and shall see in the next chapters, have been an object of profound scrutiny by historians and sociologists alike. Indeed, these developments had different trajectories, and their accurate periodization is still debated. The 1870s witnessed intense medical research about nerves, emotions and the effects of work. This interest has earlier manifestations and it continues to this day, most clearly around the concepts of 'stress' and 'burnout'. Forms of regulation of workers' leisure can be traced back at least to the early nineteenth century, and full right for annual holiday with pay was only granted in 1938. Likewise, plebian seaside recreation as well as sociability and entertainment in the spa culture existed also during the seventeenth century, and the holiday resort industry flourished more than ever after the Second World War. The time frame chosen for this research, between 1870 and the First World War, captures a

[127] Illouz, 'Introduction: Emodities or the Making of Emotional Commodities'; Illouz and Benger, 'Emotions and Consumption', 266.

[128] Borsay, *A History of Leisure*, 221.

moment of intersection between these various processes. It was at this time that the most significant shift in the epistemology of holidaymaking from physical cure to emotion management came about.

The interest in the history of tourism, travel, and holidaymaking has expanded dramatically over the past two decades, as has interest in the relation between emotions and capitalism and the history of consumption.[129] Nevertheless, research on the history of emotions' consumption is still in its fledgling stages. This book is an attempt to explore the popularization of a commercial leisure activity from the perspective of the history of emotions and to offer a historical case study of how economy and emotions were produced and reproduced simultaneously. Hence, the book will conclude with a reflection on how the emotional economy of Victorian and Edwardian holidaymaking can lead us to a better understanding of the entanglement of emotions, morality, and economy.

[129] For recent studies on the history of tourism see Alastair Durie, *Scotland and Tourism: The Long View, 1700–2015* (London: Routledge, 2017); Zuelow, *A History of Modern Tourism*; Diane Koenker, *Club Red: Vacation Travel and the Soviet Dream* (Ithaca: Cornell University Press, 2013); on the history of consumption, see Frank Trentmann (ed.), *The Oxford Handbook of the History of Consumption* (Oxford: Oxford University Press, 2012); Stearns, *Consumerism in World History*; Matt, *Keeping Up with the Joneses*.

1
Emotional pathologies and holidays as therapy

'The doctor is the true pioneer of modern health resorts', asserted an 1888 newspaper article titled *Health and Holidays*. In the following, the author explains that although the objective for visiting watering places had changed, medical guidance remained central to holidaymaking:

> One after another of English watering-places, from Bath, Brighton and Buxton, down to the most modern out-of-the-way nook, date their popularity from the time when their praises were sung by medical men. Need of rest and change is most usually the excuse for taking a holiday, and if we do not actually consult a physician as to where we shall go, yet we follow in the route of persons who have done so. And if the holiday be real and thoroughly enjoyable it is, no doubt, the best of physic.[1]

How did 'rest and change' become the purpose of holidaymaking? How did 'joy' become conceptualized as a kind of 'physic'? And why were doctors conceived to be the experts on the matter? The fact that physicians were advocating for holidays is not surprising in itself, considering the traditional therapeutic role of holidays and watering places. The puzzle is the correlation between the change of the clientele, as watering places were increasingly visited by the lower-middle and working classes, and the change in the objectives and qualities of the resorts, which were explicitly offering amusement and not just physical cure. This chapter focuses on the role that medical knowledge played in forging the link between workers and holidaymaking in watering places. More specifically, it will examine the relation between work, health, nature, and emotions.

Modernity, nerves, and emotions

Alain Corbin begins his monumental book *The Lure of the Sea: The Discovery of the Seaside in the Western World, 1750–1840* with these words: 'With few

[1] Anon., 'Health and Holidays (extract from The Christian Million)', *The Star Guernsey*, 8 September 1888/BNA.

The emotional economy of holidaymaking: Health, pleasure, and class in Britain, 1870–1918. Yaara Benger Alaluf, Oxford University Press (2021). DOI: 10.1093/oso/9780198866152.003.0002

exceptions, the classical period knew nothing of the attraction of seaside beaches, the emotion of bather plunging into the waves, or the pleasure of a stay at the seaside. A veil of repulsive images prevented the seaside from exercising its appeal.'[2] Generally speaking, until the eighteenth century natural landscapes were linked in the Western tradition to fear and repulsion. The colonization of the New World and the spreading domestication of lands in Europe started to change people's attitude towards nature in the sixteenth century. Now subject to conquest and subordinated to human needs to an unprecedented extent, nature became gradually less frightening. However, these needs were essentially material and many still associated wild nature with negative emotions, since 'old mythological explanations of nature faded; unproductive land was seen as ugly; and domesticated, inhabited landscape was seen as beautiful, and how it should be'.[3] By the end of the seventeenth century, some philosophers and theological scholars in France and Britain marked the beginning of a transition in the perception of nature. Natural theologians and Enlightenment naturalists ceased to see nature as an analogy between human and universe or between the physical and the spiritual and started to interpret nature as an expression of God—a meaningful spectacle.[4] However, travel still focused on art and classic culture, as demonstrated by the Grand Tour. Nature did not have an emotional appeal yet.

It is true that some ancient cultures, and especially the Romans, were very fond of natural springs and thermal waters for healing, for religious rituals, and for sensual pleasure (pleasures that were regarded as vulgar and immoral by many contemporaries as well as by modern Europeans). However, modern European bathing was not a direct descendant of this culture, for ancient bathing places were abandoned and destroyed with the fall of the Roman Empire and only started to be used again centuries later.[5] 'Taking the waters' in inland watering places in search of health was usually done in artificial facilities such as baths and pools. Only with the rise of Romanticism did uncivilized nature attract interest and curiosity that focused on the self and not on God. Going out into nature became an individual project. Romantic thought led both to the aesthetic appreciation of nature's authenticity and to a new intellectual and reflective interest in the emotional experience of the self. Increasingly, nature was being related to concepts of authenticity, harmony, personal freedom, and sensitivity, and emphasis was put on the individual experience of senses and emotion.[6] Paradoxically, in the case of

[2] Corbin, *The Lure of the Sea*, 1. [3] Bell and Lyall, *The Accelerated Sublime*, 7.

[4] Lorraine Daston, 'Attention and the Values of Nature in the Enlightenment', in Lorraine Daston and Fernando Vidal (eds), *The Moral Authority of Nature* (Chicago: University of Chicago Press, 2010).

[5] Wood, *Spa Culture and the Social History of Medicine*, 14–5.

[6] Borsay, 'Visualising the Seaside, c. 1750–1914', 189; Borsay, 'Town or Country? British Spas and the Urban–Rural Interface', 168; Bell and Lyall, *The Accelerated Sublime*, 7–8; Urry, *The Tourist Gaze*, 20; Corbin, *The Lure of the Sea*, 181–2.

sea bathing, pleasure was an outcome of the negative emotions, and above all fear, that were traditionally associated with nature. As Corbin puts it:

> Pleasure came from the whipping of the waves. The bather delighted in feeling the powerful forces of the immense ocean. Bathing among the waves was part of the aesthetics of the sublime: it involved facing the violent water, but without risk, enjoying the pretence that one could be swept under, and being struck by the full force of the waves, but without losing one's footing.[7]

During the nineteenth century, after the culturally widespread fear of nature was largely overcome, the connection between nature and positive emotions was strengthened. This occurred primarily as a reaction to rapid urbanization.[8] Indeed, given that the percentage of the population living in cities or towns rose from 20 per cent at the beginning of the century to 80 per cent at its end, open spaces were becoming scarcer.[9] However, it does not necessarily follow that popular imagination invested urban space with negative emotions. This was a feat gradually accomplished by the intellectual movement of anti-urbanism, which already towards the end of the eighteenth century criticized city life—seen as chaotic and over-stimulating—for its negative physical (e.g. pollution, dirt, noise), social (e.g. alienation, immoral behaviour, delinquency), and psychological effects (e.g. anxiety, nervousness, fatigue).[10] The belief that 'it is in towns that nervousness and nervous diseases most abound, their growth being encouraged apparently by the excitement of town-life'[11] drew on a new understanding of the nervous and emotional problems of modern life.

'Well, the fact seems to be that neurotic affections are increasing and multiplying on every hand', stated MD James Crichton-Browne in his address delivered at the opening of The Section of Psychology at the meeting of the British Medical Association in 1880. Crichton-Browne, one of the most eminent practitioners of neurology and psychiatry in the Victorian era, observed that 'the nervous temperament and diathesis are much more common in these islands than they used to be', attributing this to 'the influence of the conditions of modern life upon the

[7] Corbin, *The Lure of the Sea*, 73.

[8] Note that increasingly from the end of the eighteenth century, industrialization was shifting most of the production from the rural area to the urban sphere; see Borsay, *A History of Leisure*, 180–1.

[9] Urry, *The Tourist Gaze*, 18.

[10] Adams, *Healing with Water*, 13, 161; Borsay, *A History of Leisure*, 188; Berghoff, 'Modern Tourism and the Rise of the Consumer Society'; Urry, *The Tourist Gaze*; Urry, *Consuming Places*; Bailey, *Rational Recreation*, 20–2; Georg Simmel, 'The Metropolis and Mental Life', in Gary Bridge and Sophie Watson (eds), *The Blackwell City Reader*, pp. 103–10 (Oxford: Wiley-Blackwell, [1903]2002).

[11] James Crichton-Browne, 'Abstract of an Address Delivered at the Opening of the Section of Psychology at the Meeting of the British Medical Association at Cambridge', *The Lancet* 116, no. 2973 (1880): 292–5, here 295.

nervous system'. The whole text reflects a feeling of panic and fear. '[N]early 70,000 deaths are attributed to nervous disease in England and Wales each year',[12] he noted. Judging by the numbers, one might have thought Britain was facing a deadly famine.

Crichton-Browne's address encapsulates well the way many European physicians perceived the nervous disorders of their time. From the 1860s onwards, much effort was made by physicians, neuropathologists, and psychopathologists to understand the different causes and symptoms of neurasthenia, fatigue, and depression.[13] The term 'neurasthenia' (literally nerve weakness) was promoted by American neurologist George Beard in 1869, and marks the shift in scientific attention to the nerves.[14] Beard's diagnosis described neurasthenia as a variety of nervous and emotional symptoms such as neuralgia, fatigue, migraine, despair, anxiety, and depression. Although the term itself, as other terms for the same symptoms, were used by British physicians previous to Beard's work, his studies helped to popularize the diagnosis.[15] By 1900, neurasthenia had become the most common neurological diagnosis, used indiscriminately to refer to all general nervousness and psychosis. Indeed, the terms fatigue and neurasthenia were used in the eighteenth century as well, yet their late-nineteenth-century version differed in at least in two ways: first, they were now viewed in connection with new social phenomena, institutions, and 'modern life';[16] and second, they were

[12] Ibid. 293–4.

[13] Edward Shorter dubs the time between the 1860s and the First World War 'the golden years of nerves', Edward Shorter, *How Everyone Became Depressed: The Rise and Fall of the Nervous Breakdown* (Oxford University Press, 2013), 28. See also Edward Shorter, *From Paralysis to Fatigue: A History of Psychosomatic Illness in the Modern Era* (New York: The Free Press, 2008), 220–2; Petteri Pietikainen, *Neurosis and Modernity: The Age of Nervousness in Sweden* (Leiden: Brill, 2007); Marijke Gijswijt-Hofstra, 'Introduction: Cultures of Neurasthenia from Beard to the First World War', in Marijke Gijswijt-Hofstra and Roy Porter (eds), *Cultures of Neurasthenia from Beard to the First World War* (Amsterdam: Rodopi, 2001); Roy Porter, 'Nervousness, Eighteenth and Nineteenth Century Style: From Luxury to Labour', *Clio Medica* 63 (2000): 31–49.

[14] George Beard, 'Neurasthenia, or Nervous Exhaustion', *The Boston Medical and Surgical Journal* 80, no. 13 (1869): 217–21; George Miller Beard, *American Nervousness, Its Causes and Consequences: A Supplement to Nervous Exhaustion (Neurasthenia)* (New York: Putnam, 1881). See also Shorter, *The Rise and Fall of the Nervous Breakdown*, 37–44. Sigmund Freud was also deeply interested in neurasthenia and published in 1894 a paper in which he argued that the disorder should be considered an independent syndrome. However, Freud's works were generally coolly received in Great Britain before the First World War and many members of the British medical profession were quite hostile to his ideas. Therefore, the chapter will be based on British sources and on Beard's work, which was more influential in Britain. On Freud and neurasthenia, see German E. Berrios, *The History of Mental Symptoms: Descriptive Psychopathology Since the Nineteenth Century* (Cambridge: Cambridge University Press, 1996), 270–1. On Freud and the British medical community see Shorter, *From Paralysis to Fatigue*, 304–5; Dean Rapp, 'The Early Discovery of Freud by the British General Educated Public, 1912–1919', *Social History of Medicine* 3, no. 2 (1990): 217–43.

[15] Shorter, *The Rise and Fall of the Nervous Breakdown*, 37.

[16] See also Tracey Loughran, 'Hysteria and Neurasthenia in Pre-1914 British Medical Discourse and in Histories of Shell-Shock', *History of Psychiatry* 19, no. 1 (2008): 25–46, here 39.

increasingly associated with emotions and framed as emotional pathologies. In the words of historian Jan E. Goldstein, they represented and contributed to the 'medicalization of emotional life'.[17]

Beard himself acknowledged that the pathological causes of neurasthenia were not clear, but stressed that the disorder was an outcome of modern life in America, which was increasing cerebral tension and weakening the nerves.[18] Most experts on nervous disorders across the Atlantic adopted the same idea, pointing to urbanization, motorization, and technology as the major factors intensifying nervous activity.[19] They attributed the causes of 'neurotic affections' to the increasing pressure and complexity of modern life, which were believed to directly influence the nervous system and the brain.[20] This 1886 article published in the *British Medical Journal* under the title 'The Wear and Tear of Modern Life' sums up this view nicely:

> [T]here seems no reason to dispute the prevalent impression that nervous diseases are commoner now than formerly, and that we suffer vastly more than our forefathers from the sense of pressure, worry, and unrest ... This high tension, though no doubt peculiarly characteristic of metropolitan life, prevails more or less in all our large centres of population, and we cannot regard its possible further increase without apprehension, both for ourselves and for those who will come after us. To a certain extent, it is the inevitable outcome of modern civilization [sic].[21]

In addition to its intrinsic connection to modernity, research on nervous disorders went hand in hand with the growing interest in emotions and was related to the broader distinction between the psyche and the soma that was evolving during the nineteenth century. Moreover, as emotions were frequently conceptualized as

[17] Jan E. Goldstein, *Console and Classify: The French Psychiatric Profession in the Nineteenth Century* (Cambridge: Cambridge University Press, 1987), 374.

[18] Francis G. Gosling and Joyce M. Ray, 'The Right to be Sick: American Physicians and Nervous Patients, 1885–1910', *Journal of Social History* 20, no. 2 (1986): 251–67, here 252.

[19] On the medical discourse in other European countries see for instance Pietikainen, *Neurosis and Modernity*; Laura Goering, '"Russian Nervousness": Neurasthenia and National Identity in Nineteenth-Century Russia', *Medical History* 47, no. 1 (2003): 23–46; Doris Kaufmann, 'Neurasthenia in Wilhelmine Germany: Culture, Sexuality, and the Demands of Nature', in Marijke Gijswijt-Hofstra and Roy Porter (eds), *Cultures of Neurasthenia From Beard to the First World War* (Amsterdam: Rodopi, 2001); Joachim Radkau, *Das Zeitalter der Nervosität: Deutschland zwischen Bismarck und Hitler* (Munich: Carl Hanser, 1998).

[20] Crichton-Browne, 'Abstract of an Address Delivered at the Opening of the Section of Psychology at the Meeting of the British Medical Association at Cambridge' (1880), 294–5; David Yellowlees, 'An Address Delivered at the Opening of the Section of Psychology: The Causes and Prevention of Insanity', *British Medical Journal* 2, no. 1283 (1885): 204–6, here 204.

[21] Anon., 'The Wear and Tear of Modern Life', *British Medical Journal* 1, no. 1318 (1886): 650–1.

being mediated between the mental and the physical through the nervous system—an intermediate category in itself—the medical literature struggled to draw a line between emotions and nerves.[22] Throughout the century, we can identify various classifications of the emotional as a mental phenomenon, as a physical one, or as a hybrid category, as implied in the contemporaneous notions 'physiological psychology', 'the physiology of mind', or the 'psychology of the nervous system'.[23] It follows, on the one hand, that unlike passions, emotions were not seen as being intrinsically negative or pathological, but rather as a natural part of the life of the normal and healthy human; on the other hand, this implied that emotions bore the potential for abnormalities and disease, which the medical profession had the prerogative to define and diagnose and which individuals were supposed to be aware of and avoid.[24]

Indeed, throughout the nineteenth century, scientific research on the nervous system and the emotions led to the pathologization of phenomena that were not necessarily classified as pathological a century earlier. This was the fate of hysteria, neurasthenia, melancholia, and schizophrenia, to only mention a few examples. Moreover, as Rick Rylance notes, the scientific vocabulary itself was being pathologized in the last decades of the century, as the terms 'psychosis' and 'neurosis', which originally meant literally 'state of mind' and 'state of the nerves', became, tautologically, names of diseases.[25] Beard argued that the symptoms of some nervous disorders could be aggravated by negative emotions such as worry and asserted, more generally, that emotional experience might incite sickness.[26] Physicians increasingly claimed that nervous disorders derived from emotional experiences associated with modernity, such as anxiety and pressure related to the constant 'struggle for existence'.[27] More specifically, these emotional states were said to be triggered by 'annoyances of duty', 'business cares, excess of work', and

[22] Fay Bound Alberti, 'Bodies, Hearts, and Minds: Why Emotions Matter to Historians of Science and Medicine', *Isis* 100, no. 4 (2009): 798–810, here 802, 808; Dixon, *From Passions to Emotion*, 23, 45.

[23] Alberti, 'Why Emotions Matter to Historians of Science and Medicine', 802; Thomas Dixon, 'The Psychology of the Emotions in Britain and America in the Nineteenth Century: The Role of Religious and Antireligious Commitments', *Osiris 16* (2001): 288–320, here 293, 299; Danziger, *Naming the Mind*, 53; Athena Vrettos, 'From Neurosis to Narrative: The Private Life of the Nerves in "Villette and Daniel Deronda"', *Victorian Studies* 33, no. 4 (1990): 551–79, here 554; Leon Stephen Jacyna, 'The Physiology of Mind, the Unity of Nature, and the Moral Order in Victorian Thought', *The British Society for the History of Science* 14, no. 2 (1981): 109–32.

[24] See also Nikolas Rose, *The Psychological Complex: Psychology, Politics and Society in England, 1869–1939* (London: Routledge, 1985), 1–10.

[25] Rylance, *Victorian Psychology and British Culture 1850–1880*, 113.

[26] George M. Beard, *Practical Treatise on Nervous Exhaustion (Neurasthenia): Its Symptoms, Nature, Sequences, Treatment* (New York: E.B. Treat, 1899), 189–90, 44.

[27] See, for instance Crichton-Browne, 'Abstract of an Address Delivered at the Opening of the Section of Psychology at the Meeting of the British Medical Association at Cambridge' (1880), 294–95; George Johnson, 'Lectures on Some Nervous Disorders that Result from Overwork and Mental Anxiety: Lecture I', *The Lancet* 106, no. 2707 (1875): 85–7, here 85; Edwin Ash, *Mind and Health: The Mental Factor and Suggestion in Treatment, with Special Reference to Neurasthenia and Other Common Nervous Disorders* (London: H. J. Glaisher, 1910), 79–80; Adrien A. Proust and Gilbert Ballet, *The Treatment of Neurasthenia*, trans. Peter Campbell Smith (London: Henry Kimpton, 1902), 29.

'preoccupations of ambition'.[28] In his book on neurasthenia, neurologist MD Thomas Stretch Dowse went so far as to state: 'I have seen worry kill as surely and as positively as a consumption kills.'[29]

From roughly 1870 up to the 1920s, professional views on the aetiology of neurasthenia shifted from the somatic to the psychological sphere. Yet, by late nineteenth century, it was still generally perceived to be a somatic condition, despite its correlation with emotional states. The medical research was undecided about the organic causes of nervous conditions, but their physical symptoms seemed to distinguish them from mental disorders. Arguing for somatic explanations helped to avoid the 'mental illness' label, especially in a time when there was no institutionalized discipline that could account for a person's incapacity to cope with emotional conditions or difficulties with emotion management without understanding these inabilities as a form of insanity.[30] For most of the century, Victorian physicians considered the effects of negative emotions as dependent on the physical state of the nerves: tense or loose, excited or relaxed. It was only towards the end of the century that some of the emotional symptoms of these pathologies attained an independent status as a defined pathology.[31] This reflected a dramatic shift in the medical perception of the relationship between body and mind: physical symptoms were increasingly believed to be caused by mental or emotional factors.

This shift was acknowledged by physicians at the time, and it was quite problematic precisely because there was not yet an established discipline that could deal with it. In his 1875 series of lectures on overwork and mental anxiety, MD George Johnson, Professor of Medicine at King's College, had to explain to his audience that the study of psychology is scientific and is not based on theology or metaphysics.[32] Even as late as 1906, MD David Drummond, a senior physician at the Royal Victoria Hospital in Newcastle, lamented while giving a lecture on

[28] Anon., 'Holiday Seekers', *The Lancet* 2, no. 3651 (1893): 448; Proust and Ballet, *The Treatment of Neurasthenia* (1902), 29.

[29] Thomas S. Dowse, *On Brain and Nerve Exhaustion. 'Neurasthenia.' Its Nature and Curative Treatment. A Paper Read before the Medical Society of London* (London: Bailliere, Tindall & Cox, 1880), 47.

[30] Shorter, *The Rise and Fall of the Nervous Breakdown*, 8; Loughran, 'Hysteria and Neurasthenia in Pre-1914 British Medical Discourse', 29–30; Mathew Thomson, 'Neurasthenia in Britain: An Overview', in Marijke Gijswijt-Hofstra and Roy Porter (eds), *Cultures of Neurasthenia from Beard to the First World War* (Amsterdam: Rodopi, 2001).

[31] See also Shorter, *The Rise and Fall of the Nervous Breakdown*, 18; Loughran, 'Hysteria Neurasthenia in Pre-1914 British Medical Discourse'; Michael Neve, 'Public Views of Neurasthenia: Britain, 1880–1930', in Marijke Gijswijt-Hofstra and Roy Porter (eds), *Cultures of Neurasthenia from Beard to the First World War* (Amsterdam: Rodopi, 2001); Berrios, *The History of Mental Symptoms*; Vrettos, 'From Neurosis to Narrative', 554.

[32] Johnson, 'Lectures on Some Nervous Disorders that Result from Overwork and Mental Anxiety: Lecture I' (1875), 85.

neurasthenia that 'not one out of a hundred medical men knows anything of even the elements of psychology'.[33] The research and treatment of nervous disorders marked a crucial step in the evolution of psychology as an independent discipline. Towards the turn of the century, psychology developed new therapeutic techniques that were not confined to the psychical aspects of the condition reported. Most notable of these techniques was, of course, psychoanalysis. In Britain, the first journal dedicated to psychology was published in 1876, and research departments of psychology and psychological societies were established around 1900.

Pathologizing the worker

In seeking to account for something that was not insanity, but a debility caused by ongoing mental and emotional distress, the research of neurasthenia was applied, as sociologist Nikolas Rose shows, in factories, in schools, in the army, and in other institutions that had an incentive to prevent disruptions to their smooth functioning.[34] By the last decades of the century, the medical community had adopted a new concept: 'overwork', a sub-category of neurasthenia concerned only with the effects of work. Thus, at the same time that industrialization gave people occasion to coin the metaphor of the 'human motor', the medical literature was coming to be dominated by new preoccupations with the physical and mental consequences of work.[35] Human motors had a limit. Overwork, however, was never clearly defined, being conceived as both physical and mental, both the cause of nervous disorders, such as epilepsy or delirium, and an independent disorder unto itself.[36] The sources of overwork also varied and sometimes contradicted each other. During the 1870s, for example, doctors attributed the cause to physical or mental activities, intensive 'brain work', such as reading, and even to a lack of stimulus.[37]

The British medical community debated whether overwork should be considered as a contemporary public health emergency, like neurasthenia. Despite the incoherency of the concept, some believed that overwork was a deadly condition.

[33] David Drummond, 'Neurasthenia: Its Nature and Treatment', *British Medical Journal* 2, no. 2375 (1906): 11–4, here 11.

[34] Rose, *The Psychological Complex*, 180.

[35] Anson Rabinbach, *The Human Motor: Energy, Fatigue, and the Origins of Modernity* (Berkeley: University of California Press, 1992).

[36] George Johnson, 'Lectures on Some Nervous Disorders that Result from Overwork and Mental Anxiety: Lecture III', *The Lancet* 106, no. 2723 (1875): 651–4, here 651, 653; Anon., 'Overwork', *The Derby Mercury*, 5 June 1878/BNA; George H. Savage, 'Discussion on Neurasthenia and Its Treatment', *British Medical Journal* 2, no. 1758 (1894): 522–5, here 524; Robert Farquharson, 'On Overwork', *The Lancet* 107, no. 2731 (1876): 9–10.

[37] George H. Savage, 'Overwork as a Cause of Insanity', *The Lancet* 106, no. 2708 (1876): 127; Johnson, 'Lectures on Some Nervous Disorders that Result from Overwork and Mental Anxiety: Lecture I' (1875), 86; Farquharson, 'On Overwork' (1876), 9.

'Our forefathers had a saying to the effect that "No man ever died from overwork"', wrote an anonymous contributor to *The Derby Mercury*. However, he continued, 'the question has arisen whether modern habits have not deprived this "old saw" of its especial value'.[38] Likewise, in a paper read before the Medical Society of London, Dowse explained:

> Exhaustion from overwork, from undue excitement, and from worry may be considered as essentially attributes of the present age, and they are more prevalent now, than they were at any other period in the world's history. Life at high-pressure is the prominent feature of the nineteenth century, and we cannot be surprised when we find that the so-called nervous diseases and exhaustions, dipsomania and insanity, are increasing beyond all proportion to the rapid increase of the population.[39]

Nonetheless, some physicians did not consider overwork to be a direct cause of neurasthenia and were not even convinced that overwork could be regarded as an independent condition.[40] In 1894, the *British Medical Journal* dedicated a whole article to criticizing an address given by German neurologist Wilhelm Heinrich Erb on the causes of nervous disorders. The authors were not convinced by Erb's argument that there was a necessary correlation between the increasing complexity of modern life and nervousness, and in particular doubted whether overwork can be said to be a direct cause of neuroses. Moreover, and in contrast to many other scientists, they concluded that they 'have still greater doubt as to there being any amount of overwork in England'.[41] Regardless of such debates, overwork became a hot topic in British medical literature.[42]

The research on overwork paid little attention to physical injuries caused at work, emphasizing instead the nervous and emotional burden imposed on people by work, as the condition was believed to have its origins in emotionally impactful experiences. Beginning in the 1880s, drawing a correlation between overwork and the negative emotional effects of work came to be a dominant trend in the literature. Thus, overwork was considered to derive from work-related anxiety and irritation and from mental depression caused by the demands of professional life or monetary concerns.[43] Curiously, even Erb's critics, while denying the risk of

[38] Anon., 'Overwork' (1878). [39] Dowse, *On Brain and Nerve Exhaustion* (1880), 40.

[40] Savage, 'Discussion on Neurasthenia and Its Treatment' (1894), 522, 525; Anon., 'The Increasing Nervousness of Our Time', *British Medical Journal* 2, no. 1774 (1894): 1493.

[41] Anon., 'The Increasing Nervousness of Our Time' (1894), 1493.

[42] In *The Lancet* alone, articles containing the term 'overwork' in the title rose from none between 1850 and 1860 and one between 1860 and 1870 to 10 between 1870 and 1880 and decreased to 4 between 1880 and 1890, 3 between 1890 and 1900, 1 between 1900 and 1910 and none in the following two decades.

[43] Johnson, 'Lectures on Some Nervous Disorders that Result from Overwork and Mental Anxiety: Lecture III' (1875), 651, 653; Anon., 'The Wear and Tear of Modern Life' (1886), 650; Anon., 'Climate

overwork, noted that '[w]orry, not work, is the danger'.[44] Overwork was not only thought to originate from emotional states, but was also seen as generating emotional symptoms such as distress, irritability, depression, and sloth.[45] In sum, like other contemporary nervous conditions, overwork was increasingly framed as an emotional pathology. It therefore exemplifies well what Sabine Donauer described as 'desomatization'—a shift from a physiological to a psychological understanding of the experience of work.[46]

Through its focus on the emotional effects of work, the research on overwork was, though perhaps unwittingly, converting 'work' into an independent category that collided with other categories, such as class and gender. The condition of neurasthenia was commonly considered to affect members of the upper classes, who were believed to have a more developed nervous system and thus a higher degree of sensitivity than members of the lower classes. In this sense, the discourse on nervousness was inherently paradoxical, as it represented weakness and degeneration as well as sophistication and cultural cultivation.[47] Beard himself supported the notion that neurasthenia was most suffered by the intellectual elite—'brain-workers' in his words—who were more prone to nervous affections than the less developed manual labourers.[48]

While this conception continued to be held by contemporaneous intellectuals, in the last decades of the century the medical profession was gradually diagnosing cases of neurasthenia among the lower classes too. It was the concept of 'overwork' that helped transform neurasthenia into a cross-class syndrome rather than an elite disease.[49] Addressing the question of overwork, authors frequently chose very general and all-inclusive expressions such as 'the working man', 'the labour question', 'this too busy generation of working men of all grades and callings', or 'professional men'.[50] Additionally, although overwork was a vague enough concept to fit different kinds of work, the medical community was undecided

and Nerves', *The Bath Chronicle*, 19 August 1897, 6/BNA; Savage, 'Overwork as a Cause of Insanity' (1876), 127.

[44] Anon., 'The Increasing Nervousness of Our Time' (1894), 1493.

[45] Johnson, 'Lectures on Some Nervous Disorders that Result from Overwork and Mental Anxiety: Lecture III' (1875), 653; Farquharson, 'On Overwork' (1876), 9; Anon., 'Climate and Nerves' (1897).

[46] Referring to turn of the century Germany, Donauer, *Emotions at Work—Working on Emotions*.

[47] Daniel Morat, 'Urban Soundscape and Acoustic Innervation Around 1900', in Robert Beck, Ulrike Krampl, and Emmanuelle Reaillaud (eds), *Les cinq sens de la ville. Du moyen âge à nos jours* (Paris: Presses Universitaires François Rabelais, 2013), 81–2.

[48] Beard, *Practical Treatise on Nervous Exhaustion* (1899), 26.

[49] See also Gijswijt-Hofstra, 'Introduction: Cultures of Neurasthenia from Beard to the First World War', 5; Chandak Sengoopta, 'A Mob of Incoherent Symptoms? Neurasthenia in British Medical Discourse, 1860–1920', in Marijke Gijswijt-Hofstra and Roy Porter (eds), *Cultures of Neurasthenia From Beard to the First World War* (Amsterdam: Rodopi, 2001), 101–2; Shorter, *From Paralysis to Fatigue*, 224–6; Pietikainen, *Neurosis and Modernity*, 101–5; Goldstein, *Console and Classify*, 335–6; Gosling and Ray, 'The Right to Be Sick', 252–3.

[50] Anon., 'Health in Relation to Work', *The Lancet* 2, no. 3498 (1890): 570–1, here 571; Anon., 'Holiday Seekers' (1893), 448.

about this 'democratic' approach; thus, in many cases, the literature reflects the physicians' own middle-class values. Perhaps it was easy for doctors to follow Beard's focus on 'mental workers' (e.g. clerks, architects, teachers, students, artists) and the intellectual discourse on the nervous sensibility of the elites and attribute the condition mostly to members of the middle and upper classes, because this enabled the physicians themselves to identify as victims of overwork.[51]

The middle-class bias of the doctors simply failed to encompass all of the cases diagnosed, showing that overwork was affecting all types of workers: manual, mental, skilled or unskilled, in the factory, the household, the university or the office.[52] It also failed to account for the wider data that suggested a correlation between nervous disorders and the degree of the patient's 'struggle for existence', being that 'statisticians tell us that annuitants, clergymen, and the well-to-do classes who have to take no thought for the morrow, live longer than shopkeepers, artisans, and labourers, who have to contend for daily bread'.[53] By the beginning of the twentieth century, physicians explicitly declared that the victims of both neurasthenia and overwork came from all social classes and occupations.[54]

The gendered aspects of the diagnosis of and discourse on overwork are somewhat more complicated to trace. This is because common theories on women's health and nervousness were being challenged both through the medical discourse on emotions and as a result of the increasing participation of women in the work force. The literature on gender and health in the nineteenth century has established that women were more likely to be attributed with madness, irrationality, nervousness, and sickness in general.[55] Hysteria was the most popular woman-specific diagnosis, and as a type of nervousness with somatic symptoms it was typologically very close to overwork and neurasthenia. The strongly gendered perception of the disease led physicians to invest much effort in distinguishing between neurasthenia and hysteria. Opinions regarding the differences between neurasthenia and hysteria varied significantly. While some saw them as entirely different diagnoses, others considered hysteria to be a type of

[51] Sinclair W. Thomson, 'On Holiday-Making', *The Lancet* 120, no. 3091 (1882): 912; Anon., 'Holiday Seekers' (1893), 448.

[52] Savage, 'Overwork as a Cause of Insanity' (1876), 127; Yellowlees, 'An Address Delivered at the Opening of the Section of Psychology: The Causes and Prevention of Insanity' (1885), 204. See also Gosling and Ray, 'The Right to Be Sick', 253; Gijswijt-Hofstra, 'Introduction: Cultures of Neurasthenia from Beard to the First World War', 5; Sengoopta 'Neurasthenia in British Medical Discourse, 1860–1920', 101–2; Shorter, *From Paralysis to Fatigue*, 224–6.

[53] Crichton-Browne, 'Abstract of an Address Delivered at the Opening of the Section of Psychology at the Meeting of the British Medical Association at Cambridge' (1880), 295.

[54] Michell J. Clarke, *Hysteria and Neurasthenia* (London: John Lane The Bodley Head, 1905), 175.

[55] See, for instance Mary Orr Johnson, 'The Insane in 19th Century Britain: A Statistical Analysis of a Scottish Insane Asylum', *Historical Social Research* 17, no. 3 (1992): 3–20; Gosling and Ray, 'The Right to Be Sick', 257–9; Elaine Showalter, *The Female Malady: Women, Madness, and Culture in England, 1830–1980* (New York: Pantheon Books, 1986); Barbara Ehrenreich and Deirdre English, *For Her Own Good: Two Centuries of the Experts' Advice to Women* (New York: Anchor Books, 1978), 147.

neurasthenia or vice versa.[56] To complicate things further, some physicians argued for an independent diagnostic category named 'hystero-neurasthenia'.[57]

Although some physicians simply perceived of neurasthenia as the male equivalent of hysteria, in most cases gender lines did not provide a definite solution, as symptoms of hysteria were reported in men and those of neurasthenia in women.[58] Thus, MD George Johnson of King's College stressed that hysteria's 'characteristic symptoms may be fully developed, not only in those who have a healthy uterus, but in those who do not possess a uterus'. Yet, he believed that women were more prone to suffer from the condition because they had a more sensitive nervous system.[59] This was also George Beard's opinion, as he explained in his *Practical Treatise on Nervous Exhaustion*:

> Hysteria is found usually in those whose emotional natures greatly predominate. Hence, relatively to neurasthenia, it is far more common in females than in males. Indeed, hysteria was once supposed to be exclusively a disease of women; hence its name. Neurasthenia, on the other hand, although more frequent in women, is yet found in great abundance in both sexes, and both men and women of intellect, education, and well-balanced mental organizations.[60]

Despite the effort to create an analytical distinction between hysteria and neurasthenia, prejudices about women influenced research on female nervous disorders. Maintaining 'that feminine and nervous were synonymous terms',[61] physicians expected women to be the main victims of the nervousness crisis and, even up through the turn of the century, continued holding to theories that drew a causal relation between nervousness and female genitalia.

The specific condition of overwork contributed to the confusion by adding 'work' to the already overly complex gender-nervousness formula. It was generally agreed, both in Europe as in the United States, that neurasthenia was more common among men, and the linkage to work helped to mark the sub-category

[56] George Burford and Clarence Granville Hey, *Practical Summary of Points to Be Noted in Cases of Neurasthenia in Women, Including Clinical History & Physical Finding* (London: J. Bale & Co., 1907), 13; Savage, 'Discussion on Neurasthenia and Its Treatment' (1894), 525; Anon., 'The Increasing Nervousness of Our Time' (1894), 1493; Johnson, 'Lectures on Some Nervous Disorders that Result from Overwork and Mental Anxiety: Lecture III' (1875), 653. See also Sengoopta, 'Neurasthenia in British Medical Discourse, 1860–1920', 99–100.

[57] Loughran, 'Hysteria and Neurasthenia in Pre-1914 British Medical Discourse', 37.

[58] Gijswijt-Hofstra, 'Introduction: Cultures of Neurasthenia from Beard to the First World War', 7; Sengoopta, 'Neurasthenia in British Medical Discourse, 1860–1920', 97–8; Shorter, *From Paralysis to Fatigue*, 220–2.

[59] Johnson, 'Lectures on Some Nervous Disorders that Result from Overwork and Mental Anxiety: Lecture III' (1875), 653.

[60] Beard, *Practical Treatise on Nervous Exhaustion* (1899), 137.

[61] Crichton-Browne, 'Abstract of an Address Delivered at the Opening of the Section of Psychology at the Meeting of the British Medical Association at Cambridge' (1880), 294; see also Gosling and Ray, 'The Right to be Sick', 256–7.

of overwork as a principally male condition. However, as women's participation in the labour force was increasing both in industry as well as in domestic service, the effect of the gender variable needed to be reconsidered. Therefore, some physicians suggested that the effect of gender was in fact mediated through work. In their influential book on neurasthenia, the French scientists Adrien Proust and Gilbert Ballet made this point very clear. They claimed that the condition, indeed, 'is commoner among men than among women', but stressed that the 'unequal division of the disease between the sexes cannot be explained by a special predisposition on the part of the male sex. Here again appears the preponderating influence of stubborn labour, of cares, and of excesses of all sorts, consequences of the more active and more militant part played by man in the struggle for existence.'[62]

This helps clarify why lower-class women were frequently diagnosed with neurasthenia and overwork, while middle- and upper-class women were rarely considered to suffer from these specific conditions. To explain the symptoms experienced by nervous middle-class women, physicians preferred to assume that they were manifestations of gynaecological disorders. But women who had heavy responsibilities were granted with the right to be overworked, just as were middle-class men.[63] One neurologist explained that women's 'sphere of work is the home in which they sleep and take their meals', and thus while a '[m]an works from dawn to set of sun; a woman's work is never done'. This led women to be as overworked as working men.[64] This was all the more true as women were increasingly working outside of the household.

Regulating working time, regulating emotions

The British medical community was as inconclusive about the cure for overwork as it was in defining the syndrome itself. Just as it was not clear whether overwork was a diagnosis, a symptom, or a source of other diseases, the articulation of the potential cure varied from medicinal treatment, for example with opium or chloral, to electrical treatment or hypnotism, to general behavioural advice. The most common opinion was that the adequate remedy was simply to work less. However, many thought that the reduction of working time was not enough, and physicians had to ensure that in their free time workers engaged in healthy and emotionally positive activities. Thus, leisure came into the arsenal of therapeutic methods.

[62] Proust and Ballet, *The Treatment of Neurasthenia* (1902), 8.
[63] Gosling and Ray, 'The Right to Be Sick', 258–9.
[64] Charles D. Musgrove, *Holidays and How to Use Them* (Bristol: J.W. Arrowsmith, 1914), 16.

Physicians made sure to be very cautious so as not to appear that they doubted the value or necessity of work. In fact, they were caught in something of a bind. While doctors wished to encourage those suffering from overwork and explain to their relatives and employers that their condition was real, they also had to emphasize that 'lazy, idle, and indifferent men do not, as a rule, suffer from the disease of which we are writing'.[65] In the process, the meaning of work became ambiguous. Was work the cause of the diagnosis or was it worker's inadequate emotion management? Doctors understood work both as the source of brain exhaustion and as the place where people had to learn how to manage their emotions. What the medical literature concerned with overwork was calling for was, therefore, not to abolish work nor even to change the structure of labour, but merely to reduce its length.

This paved the way for physicians to stake out positions in the ongoing political and moral debates on working hours and free time. 'If the ministers of health, the professors of medical science and art, do not understand the powers of the human body in regard to the mental and physical work which it can safely perform, who do know?' said MD Benjamin Ward Richardson in a lecture delivered before the Sanitary Institute in 1890, and continued by asking his audience: 'Has our knowledge advanced to the point when we can say to the legislator, to the employer, and to the employed, so much per day can be measured out as compatible with health and natural vitality, and no more?'[66] He answered in the positive. Richardson went on to state that eight working hours should be a general standard and that every worker 'ought to have his holiday, like other persons, when he required it for his health's sake, or for his relief from the cares and worry of life'. In order to find the appropriate balance between the concern for each individual worker's health and the general community's life and business, Richardson called for stronger cooperation between legislators and physicians.[67]

Indeed, physicians were trying to shape the legislation process by presenting petitions to parliament, by appearing in front of parliamentary committees, and by writing about the issue in both general and scientific mediums. Thus, for example, *The Lancet* published a report after the passage of the Weekly Half-Holiday for Shops Bill of 1888, which for its part followed the Shop Hours Regulation Act. The report supported the reduction of daily working hours in the shop service and mentioned the effect that 'the opinion strongly expressed by influential members of our profession' had exerted on the legislators. This opinion was that long working hours posed grievous dangers to health, above all for women. The report stressed that 'the scandal of overwork now prevalent' was

[65] Dowse, *On Brain and Nerve Exhaustion* (1880), 42.

[66] Anon., 'Health in Relation to Work' (1890), 571.

[67] Ibid. 570–1.

acknowledged by all medical practitioners, but that an effective remedy required legislative intervention.[68] Discussions on questions of policy and legislation on the pages of scientific medical journals continued into the twentieth century. A decade after the passage of the Weekly Half-Holiday for Shops Bill, a contributor to *The Lancet* asserted that 'the gain in point of health both to employers and employed which the cessation of work between Thursday and Tuesday implies is considerable'.[69]

Nonetheless, the medical literature viewed securing time off from work as only part of the cure, or as a necessary condition for the cure itself. After setting the now scientifically established claim that work should be balanced with rest and recreation, the next step was to define which kind of free-time activities were beneficial. The most common opinion was that the adequate response to overwork is rest.[70] Although in many ways this approach was based on the 'rest cure' developed in the 1870s by American neurologist MD Silas Weir Mitchell, when applied to cases of overwork it usually took a substantially different form. Mitchell's method was used to treat cases of neurasthenia, hypochondria, and hysteria, and consisted of 'mechanical rest-stillness': patients were put in bed, fed mostly fluids, and required to refrain from movement. In some cases, talk and laughter were forbidden as well.[71] This coincided with Mitchell's understanding of nervous diseases and mental illnesses as essentially somatic.[72] Mitchell was very sceptical towards the new discipline of psychology. He focused mainly on the patients' physical symptoms, such as anaemia, low temperature, or fast heart rate, and believed that emotions were generally a manifestation of pain that should be relieved by rest: 'I have no doubt that very many cases of hysteria, neurasthenia with obsessions and the like have been cured by simple country doctors using full rest treatment and nothing else; *You cure the body and somehow find that the mind also is cured*.'[73]

However, as described above, such rationalizations of the relation between emotions and nervous disorders were heavily contested and Mitchell's was certainly not the only or even the most dominant view. Many physicians hypothesized the existence of a reciprocal relationship between the mental and the

[68] Anon., 'The Weekly Half-Holiday Bill' (1890), 668–69.

[69] Anon., 'The Easter Holidays Extension', *The Lancet* 1, no. 3996 (1900): 952.

[70] Savage, 'Discussion on Neurasthenia and Its Treatment' (1894), 523–4; Johnson, 'Lectures on Some Nervous Disorders that Result from Overwork and Mental Anxiety: Lecture I' (1875), 85; John Woodrow Presley, 'Neurasthenia and the Cure of Literature: Robert Graves, Siegfried Sassoon, Andy Collins', *JAC* 30, no. 1/2 (2010): 269–313, here 272.

[71] Weir S. Mitchell, 'The Treatment by Rest, Seclusion, etc. in Relation to Psychotherapy', *Journal of the American Medical Association* 50, no. 25 (1908): 2033–7, here 2035; Weir S. Mitchell, *Rest in the Treatment of Nervous Disease* (New York: Putnam's Sons, 1875), 88–90. See also Adams, *Healing with Water*, 163.

[72] Mitchell, 'The Treatment by Rest' (1908), 2036; Mitchell, *Rest in the Treatment of Nervous Disease* (1875), 96.

[73] Mitchell, 'The Treatment by Rest' (1908), 2036–7. My emphasis.

physical, while others saw emotions as the cause of nervous disorders and not the other way around.[74] Hence, ideas about the adequate therapeutic approach to treating nervous and emotional conditions varied too. MD David Yellowlees, for instance, held that 'emotions and affections are the mightiest factors in our lives, and they afford a vast field for the manifestations of nerve-instability'. Therefore, he believed that '[i]t is in the regulation of our moral nature, and in controlling our fancies, impulses, and passions, by reason and duty, that the hardest battle must be fought'. In his opinion, emotional control and management were the only way to protect the mind from the 'storms of passion' and 'the quicksands of indulgence' and thus to prevent further illnesses. More than rest and relaxation, he believed that nervous disorders could be avoided by practising self-control and emotion moderation.[75]

To paraphrase Mitchell, Yellowlees' argument was that 'you cure the mind and find out that the body is somehow also cured'. In contrast to medicinal, electrical or rest methods—all applied primarily on the patient's body—Yellowlees' words anticipated a different therapeutic method that gained popularity around the turn of the century, one that emphasized the powers of emotion management on somatic symptoms. One way to achieve emotion management was using clinical psychotherapeutic techniques, such as suggestion, but a simpler method was to encourage those suffering from overwork or nervousness to deliberately and independently manipulate their emotional state. This could be achieved, as contemporary physicians explained, through 'change'. The mission, then, was to identify the conditions affecting one's emotions and then to alter them.

Already in 1880, MD Thomas Dowse argued that in order to prevent nervous exhaustion work should be balanced with 'adequate change'.[76] MD William Wilson encouraged workers to change their mode of life, their social conditions, as well as the climate in which they lived, their thoughts, and their interests.[77] Likewise, Yellowlees advised avoiding nervous disorders by 'frequent short absences from the familiar surroundings and duties of home', and stressed that: 'An entire change, bringing new scenes, new faces, and perhaps a new language, has a wonderfully renovating power. It makes home more welcome, and familiar duties less irksome if we leave them for a time.'[78] Obtaining change required constant self-reflection, since with time every state of being could lead to negative

[74] See for instance Ash, *Mind and Health* (1910), 39, 79–80; Proust and Ballet, *The Treatment of Neurasthenia* (1902), 29, 117; Yellowlees, 'An Address Delivered at the Opening of the Section of Psychology: The Causes and Prevention of Insanity' (1885), 205.

[75] Yellowlees, 'An Address Delivered at the Opening of the Section of Psychology: The Causes and Prevention of Insanity' (1885), 205.

[76] Dowse, *On Brain and Nerve Exhaustion* (1880), 42.

[77] William S. Wilson, *The Ocean as a Health-Resort: A Handbook of Practical Information as to Sea-Voyages for the Use of Tourists and Invalids* (London: J. & A. Churchill, 1880), 7–8, 10, 13.

[78] Yellowlees, 'An Address Delivered at the Opening of the Section of Psychology: The Causes and Prevention of Insanity' (1885), 205.

emotions. Towards the turn of the century, this relativist view challenged the meaning of 'rest' held by people like Mitchell. Thus, MD Savage contended that 'rest is purely relative. What is rest to one man is not rest to another.'[79] Likewise, replying to Savage's paper, MD Ralph Browne stressed: 'If overwork or overstrain be the cause, then rest would seem to be the rational treatment. But rest is also a relative term; and the best treatment is that which gives the form of rest required in each particular case.'[80] Being a relative term, rest was incorporated into the therapeutic discourse of change and emotion management, as articulated in MD Charles Musgrove's book *Nervous Breakdowns and How to Avoid Them*:

> We can arrest the movements of the body, but it is infinitely more difficult to stop the workings of the mind. The engine is going at full speed, and we are unable to pull it up. But we can do something equally efficacious, we can switch it on to a different line. We can give it change. And change is rest.[81]

Holiday as change

This analysis of the emotional dimension of the medical views on modern life and the therapeutic theorization of change can help us better understand the medical approach to holidaymaking.[82] Physicians defined holiday as a form of change and stressed that 'an occasional holiday, with change of scene and occupation' benefited everyone's mind and body. 'This is no mere theory', it was claimed, 'it forms the universal experience of mankind.'[83] But change was indeed a theory, and as such it was also supposed to be professionally defined and regulated in order to prevent harm. '"Change" may do good', wrote MD Joseph Mortimer Granville, 'it is probably the best remedy to use, but... we need to determine precisely what form of change a particular case requires before we can intelligently prescribe it.'[84] The ideal change was said to be a symmetrical experience of contrast from the patient's daily life:

[79] Savage, 'Discussion on Neurasthenia and Its Treatment' (1894), 523.

[80] Ibid. 524.

[81] Charles D. Musgrove, *Nervous Breakdowns and How to Avoid Them* (New York: Funk and Wagnalls Company, 1913), 129.

[82] For a thorough description of the various trends and methods in the 'water cure' see Adams, *Healing with Water*.

[83] Anon., 'The Proper Holiday', *The Lancet* 158, no. 4067 (1901): 380; Anon., 'The Long Vacation', *British Medical Journal* 2, no. 710 (1874): 177.

[84] Joseph Mortimer Granville, '"Change" as a Mental Restorative', *The Lancet* 115, no. 2963 (1880): 907–9, here 907. In addition to his interest in the therapeutic benefits of leisure, Granville was the inventor of the first electrical vibrator, which is, interestingly, another method of cure through pleasure. See Rachel P. Maines, *The Technology of Orgasm: 'Hysteria', the Vibrator, and Women's Sexual Satisfaction* (Baltimore: The Johns Hopkins University Press, 1999).

> Those whose lives are sedentary should exercise themselves when on their holiday, while he who is always running about when at work will do well to lie on his back on the soft meadow or, pillowed by the shelving beach, listen to the lazy lapping of the waves... The hard reader takes his books with him, the man who is always on his legs goes away and, climbs, *few seek as recreation the opposite of their usual occupation*, and many a holiday longed for as the opportunity for renewing health and vigour turns out a disappointment.[85]

Work being a source of nervousness and exhaustion did not necessarily imply that the solution or cure of these maladies was to leave one's house and city to a holiday resort. However, the combination of anti-urbanist notions, the 'scientification' of British balneology, climatology, and climatotherapy, theories of change on the one hand and the pathologization of work on the other turned physical distancing into a logical solution.[86] That was especially so in the case of urban workers, who were thus advised to 'seek renewal of health by temporary abstention from work and departure from home'.[87] Such spatial differentiation was essentially supposed to foster a change in emotional state. Change, explained MD Granville, is 'a remedy'; but when applied by the patient it is 'not merely a change of place, of air, of scene, or of the surroundings, but a change in the inner workings of his mind, which lifts him out of a morbid groove... back on to the old lines of his life'.[88] This extract reveals a theory of emotion management achieved through a change in space. However, for a change in space to produce such an effect, spaces themselves needed to be previously associated with specific emotional states. Here we come full circle back to the shifts in the emotional charge of nature outlined at the beginning of the chapter. The support of holidaymaking by the late nineteenth-century medical community was an outcome of three historical processes: the emotionalization of nature, the emotional pathologization of modernity, urban life and work, and the therapeutic perception of emotion management. While physicians' approach to holidays in nature was formerly focused on the therapeutic effect of specific elements like minerals or ozone on physical maladies, in the medical literature of the end of the century, nature as a whole was perceived as a source of healthful emotional alteration and relief. As stated in *The Lancet*: 'In this era of universal activity and of struggling competition we are happily not

[85] Anon., 'The Proper Holiday' (1901), 380. My emphasis.

[86] The influence of climatological research on holiday practices will be addressed in the coming chapters. For an overview of British climatology from 1815 see Vladimir Jankovic, 'The Last Resort: A British Perspective on the Medical South, 1815–1870', *Journal of Intercultural Studies* 27, no. 3 (2006): 271–98; on the gradual incorporation of climatology and balneology into the authoritative medical sciences in the last decades of the nineteenth century, see Adams, *Healing with Water*, 3, 45–6; on the concept of healthy climates in relation to British seaside resorts, see Beckerson and Walton, 'Marketing the Intangible at British Resorts'.

[87] Anon., 'The Long Vacation', *British Medical Journal* 2, no. 710 (1874): 177.

[88] Granville, '"Change" As a Mental Restorative' (1880), 908.

without the consoling reflection that Nature steadily makes compensation wherever excess has occurred, and sooner or later has humanity with her in the work.'[89]

Just as the syndrome was being articulated in emotional terms so too was nature, and thus it was regarded as the ideal and healthiest destination for holidays. Physicians were not merely describing the effect of ozone on rheumatism or of sea water on gout, but were instead pointing to the holistic effect of nature itself on health. A good example can be found in MD William Squire's article on 'The Sulphur Springs of Great Britain and their Therapeutic Action', published in *The Lancet* in August 1888. Although Squire was an epidemiologist specialized in consumption, and although the article was concerned with the effects of mineral springs, it also argued that 'visits to different health resorts often prove beneficial by mere change of air, of scene, of society, amusements, or occupation'. The contribution to one's health was determined mostly by these elements, and only 'in some degree by the directly medicinal properties of their mineral waters'.[90]

Most clearly from the 1880s onward, the medical literature came to explicitly describe holidaymaking as a means to cope with the emotional burden of workers' daily routines. Holidays were seen as the ideal method to shake off for a time the worries and anxieties of daily life. Thus, we can find physicians recommending particular resorts, not because of their climactic qualities, for example, but for offering better 'relief from the pressure of an overworked and aggressive civilization' than others,[91] while their colleagues argued that 'great change' and 'open-air life' were the best remedy for nervousness, overwork and 'all worry connected with business and domestic affairs'.[92]

The pathologization of workers and the perception of holidays as a therapeutic practice makes itself most evident in the association of the terms 'invalid', 'worker', and 'overwork' with diseases. Thus, for example, it was said that the seaside 'is useful in cases of anemia, and general loss of tone induced by overwork, by chronic illnesses, or by other depressing agencies'; the ocean was claimed to 'benefit weak anemic youths of both sexes and people suffering from general debility and nervous exhaustion from overwork'; and health resorts were increasingly recommended for 'those suffering from nervous diseases' as well as for 'the jaded Londoner or tired man of business'. Holidays, it was argued, offered a way to recharge and go on; they were a 'short span of varied activity' by which workers 'have to find health, courage, and energy for another year's work'.[93]

[89] Anon., 'The Weekly Half-Holiday Bill' (1890), 668–9, here 668.

[90] W. M. Squire, 'The Sulphur Springs of Great Britain and their Therapeutic Action', *The Lancet* 132, no. (1888): 201–3, here 201.

[91] Isaac Burney Yeo, *Climate and Health Resorts* (London: Cassell & Company, 1890), 107.

[92] Hy. M. Doyle, 'Voyaging for Health: A Contribution to the Therapeutics of the Sea', *The Lancet* 136, no. 3492 (1890): 226–8, here 227.

[93] Aylmer MacFarlane, 'Some Remarks on Brighton as a Seaside Health Resort', *The Journal of Balneology and Climatology* 1, no. 3 (1897): 232–5, here 234; Doyle, 'Voyaging for Health:

The therapeutic perception of holidaymaking in watering places did not vanish during the nineteenth century. This chapter has outlined the role of the medical community in the emotional economy of holidaymaking. Medical scientists made their contribution by construing emotions as objects that can be pathologized, desired, and managed. There were more sceptical opinions about holidays, and some warned that people might actually get worse during their vacation. Yet, this concern only served to reaffirm the need for medical professionals to conceptualize and design the right and healthy holiday, because many dangers, so the theory went, could be effectuated by poorly executed holidays, poor sanitary conditions, or the over-exercising of a weak body.[94] Framed in therapeutic terms, holidaymaking was increasingly subjected to the expertise of the physician, who was allegedly the only one who could 'pause to reflect how that interval of freedom is generally spent, and whether it is always of as great benefit as it should be'.[95]

The particular ways in which the medical interest in overwork affected the emotional economy of British holidaymaking will be explored in the following chapters, but some of its cultural, political, and economic impacts can be already discerned: it enabled emotional states such as happiness, fun, and relaxation to acquire new value and social legitimation; it facilitated the expansion of formerly elite practices to lower classes through the inclusive category of 'the worker'; it reinforced the emerging *ethos* of emotional self-reflection and management, while at the same time positioning the medical profession as the scientific authority on the conceptualization of and research into emotions; and finally, it paved the way for physicians to get involved in the contemporary debate on working time as well as in the design and operation of leisure activities that were not formerly perceived as medical.

A Contribution to the Therapeutics of the Sea' (1890), 227; Arthur Oliver Holbeche, 'The Claims of British Resorts, With Some Remarks on Malvern as a Health Resort', *The Journal of Balneology and Climatology* 1, no. 4 (1897): 332–5, here 334–5.

[94] Anon., 'The Proper Holiday' (1901), 380; Anon., 'Holidays', *The Bolton Chronicle*, 16 September 1871, 5/BNA.

[95] Farquharson, 'On Overwork' (1876), 9; see also Johnson, 'Lectures on Some Nervous Disorders that Result from Overwork and Mental Anxiety: Lecture I' (1875), 86.

2
Holiday legislation as remedy

'The times have changed, and men have changed with them', declared the anonymous author of an 1889 article titled 'The Holidays'.[1] This feeling of a substantial transformation, not much different from that expressed by the medical community, echoes throughout most of the discussion on working time in the general press. Supporters of holiday legislation acknowledged the complications that such legislation might have posed, including 'the interruption of business, the dislocation of traffic on railways . . . the demoralization of labourers, and all the rest of it'.[2] They also recognized that their forefathers worked longer hours, but saw the prolongation of workers' holidays as an essential characteristic of their times, arguing that any attempt to resist this development would be destined to fail. As one writer noted, workers of previous generations 'who toiled and moiled through all the fifty-two weeks in the year without any thought of relaxation, would have been scandalized at the demands of the present generation for frequent seasons of respite from labour'.[3] Yet, those favouring holiday legislation insisted that these demands did not appear 'because the race has degenerated, or because the men's minds run more to pleasure',[4] but rather because the 'industrial and social conditions have entirely changed within fifty years'.[5] Advocates argued against those objecting to holiday legislation, proclaiming that they 'forget that with a change of times there has been a change of circumstances' that requires new coping mechanisms.[6]

But what changed exactly? What implicated this deep transformation of men, women, and society? From the commentary on holiday legislation the answer seems to be pretty clear: the effect of modern times on people's lives, and particularly on their nervous system. These changes included the high pressure of work and the intensity of factory life; the growing erosion of time and space, experienced as acceleration of movement and of communication as well as a material shortening of distances; higher competition and complexity; life in crowded cities.[7] The 'forbears laboured under less enervating circumstances . . .

[1] Anon., 'The Holidays', *The North-Eastern Daily Gazette*, 10 June 1889/BNA. [2] Ibid.

[3] Anon., 'The Development of Holidays', *The Bristol Mercury*, 27 December 1895, 6/BNA.

[4] Anon., 'Bank Holidays', *The Western Times*, 5 April 1904, 7/BNA.

[5] Anon., 'Bank Holidays and Health', *The Western Times*, 28 May 1901, 4/BNA.

[6] Anon., 'The Holidays', *The North-Eastern Daily Gazette*, 10 June 1889/BNA.

[7] Anon., 'The Development of Holidays', *The Bristol Mercury*, 27 December 1895, 6/BNA; Anon., 'Bank Holidays', *The Western Times*, 5 April 1904, 7/BNA; Anon., 'The Development of Holidays', *The*

The emotional economy of holidaymaking: Health, pleasure, and class in Britain, 1870–1918. Yaara Benger Alaluf, Oxford University Press (2021). © Yaara Benger Alaluf. DOI: 10.1093/oso/9780198866152.003.0003

[t]hey lived normal, natural lives, less complex, less exciting, less nervous, than is possible to-day', but the growing nervousness and pressure, it was argued, increased the need for periodical leisure.[8] The present chapter seeks to examine how this general sentiment materialized into formal legislation and how holidays for workers were legitimized despite strong opposition.

Towards holiday legislation

During the first half of the nineteenth century, the British working week extended significantly, and while in the eighteenth century workers generally controlled their own working hours and enjoyed their traditional holidays (e.g. St Monday, Wakes Week), industrialization and factory discipline were changing the reality of workers' lives.[9] Average working time rose from 3,000 hours a year in the mid-eighteenth century to more than 4,000 hours by the first decades of the nineteenth century, and days of holiday decreased from over 40 days in the second half of the eighteenth century to four days by 1834.[10] This change was most significant for factory workers for whom the innovation of artificial light meant working at nights and who in addition to working more were also subject to the intense process of urbanization, which eroded many familiar modes of recreation and compounded workers' experience of alienation.[11]

Beginning in the 1840s, workers started demanding their pre-industrial traditional holidays back. The most successful example is that of northern England textile workers, who managed to secure a full-week break by leaving towns en masse during the Wakes Week.[12] Gradually, these kinds of actions were institutionalized into collective bargains between organized workers and their employers. In many industrial towns, especially in northwest England, time off work was

Bristol Mercury, 27 December 1895, 6/BNA; Anon., 'The Holidays', *The North-Eastern Daily Gazette*, 10 June 1889/BNA; Anon., 'Bank Holidays and Health', *The Western Times*, 28 May 1901, 4/BNA.

[8] Anon., 'Bank Holidays and Health', *The Western Times*, 28 May 1901, 4/BNA; Anon., 'The Holidays', The *North-Eastern Daily Gazette*, 10 June 1889/BNA; Anon., 'The Development of Holidays', *The Bristol Mercury*, 27 December 1895, 6/BNA.

[9] Borsay, *A History of Leisure*, 177, 193, 201–2; Douglas A. Reid, 'Weddings, Weekdays, Work and Leisure in Urban England 1791–1911: The Decline of Saint Monday Revisited', *Past & Present* 153, no. 1 (1996): 135–62; Alan Metcalfe, *Leisure and Recreation in a Victorian Mining Community: The Social Economy of Leisure in North-East England, 1820–1914* (London: Taylor & Francis, 2005), 21; Herman Freudenberger and Gaylord Cummins, 'Health, Work, and Leisure Before the Industrial Revolution', *Explorations in Economic History* 13, no. 1 (1976): 1–12; Stearns, *Lives of Labor*, 239–40.

[10] Barton, 'Why We're All Going on a Summer Holiday', 33.

[11] Stearns, *Lives of Labor*, 5–6, 346; Albert Larking, *History of the Early Closing Association and How the Saturday Half-Holiday Was Won* (London: Unwin Brothers Ltd, 1914), 3.

[12] Walton, 'The Demand for Working-Class Seaside Holidays in Victorian England', 254–5; Poole, 'Oldham Wakes'.

standardized, specifically on Mondays and on Saturday afternoons.[13] By mid-century, these local and sporadic efforts were slowly being joined by unions as well as some members of the upper classes who advocated for shorter working hours and better recreation opportunities. Prominent representatives included the Eight Hour Day Movement and the Early Closing Association. These movements were the first to raise awareness of the problem of long working hours and led many employers to apply early closing regulations and half-holidays.[14] These customs were later on realized in official legislation, for example in the Factory Acts and Workshop Acts of the mid-century, which made Saturday half-holiday compulsory in some industries.[15]

Having some free time, a new question appeared: what could and should workers do in their time off work? In their limited free time, workers wandered the streets, gathered in small parks, or visited the pubs, simply because there were hardly any other recreational opportunities for working-class people in the fast-growing industrial cities. Some workers engaged in different types of games and races, which together with alcohol consumption were seen by the upper classes as immoral, irrational, and non-productive activities. The vulgar and unbearable presence of this loud and often violent crowd gave rise to 'the leisure problem', which called for an organized reform movement. Popular fairs, races, and some drinking habits were being repressed by the authorities.[16] However, by the middle of the century it became clear that a positive approach should be incorporated into the project of the moralization of the working masses. Employers, officials, and moralists argued that leisure time had to be rationalized and controlled just like working time.[17] After the condemnation of most of the old-fashioned amusements of the lower classes came the need to define new, acceptable forms of leisure. The writer Walter Besant expressed this mission very explicitly in his article 'The Amusements of the People': 'we have suppressed every single one of these amusements. What have we put in their place?'[18]

[13] Barton, *Working-Class Organisations and Popular Tourism*, 79–81; Reid, 'Work and Leisure in Urban England'; Bienefeld, *Working Hours in British Industry*, 71–2, 85–7.

[14] Anon., *The Beginnings of the Early Closing Movement* (London: Spottiswoode, Ballantyne & Co Ltd, printers, 1917); Larking, *History of the Early Closing Association*, 11; Bailey, *Rational Recreation*, 80.

[15] Larking, *History of the Early Closing Association*; HC Deb 1 March 1889, vol. 333 c720/PP; Reid, 'Work and Leisure in Urban England', 162; Pimlott, *The Englishman's Holiday*, 142–4; William Warrender Mackenzie Baron Amulree, 'Industrial Holidays', *Journal of the Royal Society of Arts* 87, no. 4486 (1938): 3–14, 5. Earlier attempts to launch movements for legislation usually failed: Bienefeld, *Working Hours in British Industry*, 57–8. It should be noted that early closing on Saturdays was a British phenomenon that was very rare on the continent: Huberman, 'Working Hours of the World Unite', 972.

[16] Urry, *The Tourist Gaze*, 19; Bailey, *Rational Recreation*, 19, 97–8.

[17] Lowerson and Myerscough, *Time to Spare in Victorian England*, 16; Bailey, *Rational Recreation*; Meller, *Leisure and the Changing City*, 11; Pimlott, *The Englishman's Holiday*, 85; Baker, 'The Leisure Revolution in Victorian England', 80–2; Beaven, *Leisure, Citizenship and Working-Class Men in Britain*, 16.

[18] Walter Besant, 'The Amusements of the People', *Contemporary Review* 45 (1884): 342–53, here 343.

Hence, from the mid-nineteenth century onwards, efforts to 'civilize' and 'rationalize' working-class recreational habits and put a stop to 'moral degradation' were being made by diverse philanthropists, religious, and political groups.[19] As Peter Bailey shows, in addition to their support of the struggle to reduce working hours, they sought to moralize working-class leisure by providing accessible forms of 'rational recreation': concerts, sports, lectures, men's clubs, and countryside excursions. The reformers wished at once to educate the workers such that they would choose respectable leisure activities, to reduce the violence and disorder of the burgeoning city, to prevent the mobilization of the working classes through political radicalism, and, finally, to increase workers' economic efficiency by improving their performance during working hours.[20]

While the central goal of the Eight Hour Movement and similar efforts was 'to prevent undue exhaustion of the workman's vital energies',[21] support for organized urban recreation for workers was generally grounded in moral objectives. The basic assumption was that by toiling for twelve or fifteen hours a day, workers 'not only become old while young, but become demoralized in comparison with men working under ordinary conditions'.[22] Thus, reformers maintained that time off of work should be used to educate workers to drink responsibly and become active citizens interested in social issues, public life, study, and culture.[23] Indeed, contemporary evaluations of the success of the Eight Hour Movement stated that its 'moral effect is marvellous',[24] that 'the contrast between the old regime and the new is like that between day and night, or civilization and barbarism',[25] and 'that it had made bad men good, and good men better'.[26]

[19] Meller, *Leisure and the Changing City*, 11–16; Pimlott, *The Englishman's Holiday*, 85; Baker, 'The Leisure Revolution in Victorian England', 81–2; Beaven, *Leisure, Citizenship and Working-Class Men in Britain*, 16.

[20] Bailey, *Rational Recreation*. See also Borsay, *A History of Leisure*, 90; Lowerson and Myerscough, *Time to Spare in Victorian England*, 67–78.

[21] Eight Hour Standard Working Day National Defence Found (Draft), 1899/TUC/HD 5106.

[22] Conditions in British Iron and Steel Works, A Speech Delivered to the Special Commission on Hours of Labour, International Association for Labour Legislation by Alderman P. Walls, JP, 1912/TUC/HD 5106.

[23] Walls, 'Conditions in British Iron and Steel Works' (1912)/TUC/HD 5106; A Plea for the Eight-Hours' Day in Continuous Processes, by A. H. Crosfield, Presented to the Special Commission on Hours of Work in Continuous Processes, London, 7 June 1912, 4/TUC/HD 5106; Report of the Special Commission of Hours of Labour in Continuous Industries, Presented to the 7th Delegates' Meeting of the International Association for Labour Legislation, Zurich (London: The Pioneer Press, Ltd., 1912), 4/TUC/HD 5106; Report of the Committee Appointed by the Victorian Operative Masons' Society to Inquire into the Origin of the Eight-Hours' Movement in Victoria, June 11 (Melbourne: Walker, May & Co., 1884); Bailey, *Rational Recreation*.

[24] Walls, 'Conditions in British Iron and Steel Works' (1912)/TUC/HD 5106.

[25] Crosfield, 'A Plea for the Eight-Hours' Day in Continuous Processes' (1912), 4/TUC HD 5106.

[26] Walls, 'Conditions in British Iron and Steel Works' (1912), 5/TUC/HD 5106.

Until the 1870s, reformers' efforts focused on organizing and disciplining workers' recreational activities in the few free hours they had after work. The possibility of longer holidays consisting of a number of consecutive days was still far from being a general privilege. And on the organizational level, it was an issue generally left to be settled between workers and their employers. In the 1880s and afterwards, it became increasingly common for workers who had been employed for at least one year by the same employer to get an annual paid holiday of a week and sometimes a fortnight.[27] For middle-class businessmen, what is known as 'weekend' was becoming common at the same time, and for some it was possible to 'go on a holiday from Friday night until Tuesday morning'.[28]

These schemes were generally executed under collective agreements between workers and their employers. Yet not everyone was satisfied with the bargaining system. While some feared the power of the organized workers, who allegedly 'took the law into their own hands' and just went on 'any holidays which they required',[29] others thought that under this setup the worker was left '[i]solated and helpless... absolutely at the mercy of the employer'.[30] Union activists and reformers such as Gertrude Tuckwell and Sidney Webb were sceptical that working conditions could be improved through bargaining mechanisms and warned that under the pressures of increasing competition, workers would always be at a disadvantage and would thus be forced to constantly accept poorer conditions.[31] As Peter Stearns shows, compared to France and Germany, British unions preferred legislation over the potential risks of strikes, and indeed, in the British case, legislation was, on the whole, more effective than collective bargaining.[32]

The weaknesses of the bargaining system were particularly apparent in the case of female workers, who did not have strong political leadership,[33] and in the case of lower-middle-class workers, who were, for all practical purposes, not organized at all, and certainly not before the first decade of the twentieth century.[34] Even

[27] Barton, *Working-Class Organisations and Popular Tourism*, 79; HC Deb 9 March 1882, vol. 267 c450/PP; HC Deb 10 March 1884, vol. 285 c1015/PP; Holidays of Lower Division Clerks, Treasury Minute Respecting the Holidays of Lower Division Clerks (HC C.4118, 1884), 3/PP; Report and Special Report from the Select Committee on the Shop Hours Bill. Together with the Proceedings of the Committee, Minutes of Evidence, Appendix and Index (HC 287, 1892), 4, 66/PP; Reports to the Secretary of State for the Home Department on the Shops Act, 1912, as it applies to refreshment premises, by Miss A. M. Ahrons and Mr. A. K. Clark-Kennedy, H.M. inspectors of factories (Cd. 6579, 1912), 4/PP.

[28] HC Deb 2 August 1883 vol. 282 c1344/PP.

[29] HC Deb 24 February 1875 vol. 222 c800/PP.

[30] Gertrude Tuckwell, *Women's Work and Factory Legislation: The Amending Act of 1895, A Paper Read Before the Fabian Society* (London: Co-operative Printing Society Limited, 1895), 3–4.

[31] Tuckwell, *Women's Work and Factory Legislation* (895), 5; Sidney Webb, *Women and the Factory Acts* (London: The Fabian Society, 1896), 5–7.

[32] Stearns, *Lives of Labor*, 250, 254.

[33] Tuckwell, *Women's Work and Factory Legislation* (1895), 4–5; Select Committee on the Shop Hours Bill (HC 287, 1892), 3/PP.

[34] Geoffrey Crossick, 'The Emergence of the Lower Middle Class in Britain: A Discussion', in *The Lower Middle Class in Britain, 1870–1914*, ed. Geoffrey Crossick (New York: St Martin Press, 1977), 24, 43; Christopher P. Hosgood, '"Mercantile Monasteries": Shops, Shop Assistants, and Shop Life in Late-Victorian and Edwardian Britain', *Journal of British Studies* 38, no. 3 (1999): 325–56, here 331, 347.

though their working conditions were usually worse, their working hours longer, and their income lower, because women's organizations were weaker than men's unions, it was generally accepted that women 'must look to the factory legislation for help'.[35] Similarly, lower-middle-class workers usually failed to fight collectively to improve their conditions. This was especially true of those employed in white-collar occupations, such as bank and post clerks, where holidays were believed to be impossible because of the effect they might have on other institutions and businesses.[36] Thus, in the end, it was the situation of female and clerk workers that incited the explicit demand for holiday legislation.

Throughout public debates on women's working time and holidays, the suggestion that the new regulations be extended to all adult workers, regardless of their occupation and gender, was raised frequently. Likewise, it was usually dismissed. It was the problem of male clerks that led to the first legislative regulation of holiday time for adult male workers: The Bank Holidays Act of 1871. Introduced by MP John Lubbock in February 1871 and approved in May of the same year, the Bank Holidays Act provided a Saturday half-holiday and a secular holiday on the first Monday of August for bank clerks. Although officially only applying to bank clerks, the Act paved the way for general holidays for all working people. This did not escape the legislators. While opponents in Parliament carefully raised concerns over the possibility that granting bank clerks holidays today would snowball into granting them to all workers tomorrow, some of the supporters of the bill, including Lubbock himself, admitted explicitly that achieving holidays for bank clerks was only the beginning.

A year after the bill passed, MP Auberon Herbert warned his colleagues to 'look well ahead before they entered on a course of which they could not see the end, because when they wished to stop short they might find themselves unable to do so'.[37] A few years later, MP Henry James, after clarifying that 'he did not grudge holidays to any member of the community', noted that he was very worried that '[i]f it passed, it would become highly important to know where that description of legislation was to end'.[38] These were very justified worries. Reading the minutes of the Parliamentary debates on holiday legislation from the passage of the 1871 Bank Holidays Act to the First World War, it is easy to recognize the snowball effect that the first act had. Already in the debates over the 1871 bill, some suggested the omission of the word 'bank', arguing that there was 'no reason why these opportunities of relaxation should be confined to persons employed in banks'.[39] Although '[w]hat was to be desired was that all persons in this hard-working country should have such opportunities', the suggestion to substitute the word 'bank' with the word 'general' was rejected.[40]

35 Tuckwell, *Women's Work and Factory Legislation* (1895): 4–5.

36 HL Deb 4 May 1871 vol. 206 cc128-129/PP.

37 HC Deb 16 July 1872 vol. 212 c1276/PP.

38 HC Deb 6 April 1875 vol. 223 c392/PP.

39 HL Deb 16 May 1871 vol. 206 c862/PP.

40 HL Deb 16 May 1871 vol. 206 c862/PP. See also Pimlott, *The Englishman's Holiday*, 146–7.

Although the language of the 1871 bill was limited to banks, it soon fulfilled Herbert's prophecy and was extended to other institutions. In discussions leading up to the passage of the Bank Holidays Extension Act in 1875, representatives stressed that it 'established no new principle, but merely extended the operation of an Act which was passed in the year 1871, with the almost unanimous concurrence of Parliament'.[41] Because many workers were affected by the holidays of bank clerks, they were given holidays as well: in 1875, the Act was extended to some government officers, the docks and the custom houses, and adopted de facto by many businesses, including Baltic shipping, the stock and corn exchange, insurance and railway offices, law courts, and more.[42] Furthermore, the discussions on the extensions of the original act referred also to workers who were paid by the hour.[43] During the 1880s, Parliament was faced with demands to equalize the holiday benefits of rural and town officers and of female and male clerks in some occupations.[44]

As many feared, the 1871 Bank Holidays Act started a process that was very hard to stop. The Act, which had given many employees 'the benefit of the holidays',[45] was regarded as a big success. This allowed supporters of holiday legislation, including Lubbock, to now argue that it was actually 'a misapprehension to suppose that the Act was intended to apply only to the Banks, although it was necessary to use a term which should distinguish Bank holidays from other holidays'.[46] On a different occasion, MP Charles Ritchie declared that 'bankers' clerks performed very important duties; but there were others also who had heavy responsibilities and performed equally important duties'. He therefore was certain that 'the intention of the Legislature was to include them within the scope of the Bill'.[47]

Legislation for working minors and women had existed in Britain since the Factory Act of 1844, but regulation of male adult work was new. Perhaps predictably, most of the debate in Parliament around bank holidays thus focused on the question of legal interference. The general tendency was to not intervene legally in matters of working time unless it involved a case in which workers were clearly incapable of securing their rights for themselves. 'If Parliament was to commence by stating as to the holidays of adult males employed in factories, it would soon commence to state what their hours of work should be', warned MP John Hibbert, adding that only few other issues 'could involve greater danger than an attempt to

[41] HC Deb 24 February 1875 vol. 222 c795/PP.

[42] HC Deb 24 February 1875 vol. 222 c796, cc804-805; HC Deb 6 April 1875 vol. 223 cc392-8/PP. See also Barton, *Working-Class Organisations and Popular Tourism*, 90–1; Pimlott, *The Englishman's Holiday*, 149.

[43] HC Deb 24 February 1875 vol. 222 cc806–807; HC Deb 6 April 1875 vol. 223 cc392–8/PP.

[44] HC Deb 9 March 1882 vol. 267 c450; HC Deb 10 March 1884 vol. 285 c1015; HC Deb 26 August 1886 vol. 308 cc559-60/PP.

[45] HC Deb 6 April 1875 vol. 223 c394/PP.

[46] Ibid. c395/PP.

[47] HC Deb 24 February 1875 vol. 222 c796/PP.

interfere with the hours of adult male labour'.[48] MP Herbert, a central opponent of the holiday legislation, explained that once questions of labour had been introduced 'into the House they became matters of politics and sentiment', because members of the Parliament are not able to 'stand as fair and impartial judges between labour and capital'.[49]

Alongside the argument that politicians should not interfere in economic decisions stood a more general question regarding the objectives of legislation versus the autonomy of the citizens, specifically male citizens. Some argued that legislative protection should be granted 'to those who could not protect themselves—such as children and animals, that could not escape from cruelty', while adults must be treated 'as grownup and intelligent beings', without limiting their independent judgment 'by such Legislative interference'.[50] This issue concerned not only Members of Parliament, but economists, reformers, and union activists as well. Among others, the social reformer Sidney Webb also addressed the opponents of the women's Factory Acts, who 'usually speak of legal regulation as something which, whether for men or for women, decreases personal freedom, diminishes productive capacity, and handicaps the worker in the struggle for existence'. Webb discussed at length whether state regulation would 'increase or decrease the economic independence and industrial efficiency of the workers concerned' and if 'taking the whole population and all considerations into account, the evils will be greater under legislation or under free competition', concluding that:

> In the life of the wage-earning class, absence of regulation does not mean personal freedom. Fifty years' experience shows that Factory legislation, far from dimminishing [sic] individual liberty, greatly increases the personal freedom of the workers who are subject to it. Everyone knows that the Lancashire woman weaver, whose hours of labor and conditions of work are rigidly fixed by law, enjoys, for this very reason, more personal liberty than the unregulated laundry-woman in Notting Hill. She is not only a more efficient producer, and more capable of associating with her fellows in Trade Unions, Friendly Societies, and Co-operatives Stores, but an enormously more independent and self-reliant citizen. It is the law, in fact, which is the mother of freedom.[51]

Many legislators believed that male workers (in contrast to women and children) were able to negotiate their demands independently, thus viewing legal interference as a 'destruction of the principle of self-reliance'.[52] MP Herbert found it outrageous to provide legislative protection to workers, especially as 'the

[48] HC Deb 25 February 1878 vol. 238 c314/PP. [49] HC Deb 16 July 1872 vol. 212 c1276/PP.
[50] Ibid. [51] Webb, *Women and the Factory Acts* (1896), 4–5.
[52] HC Deb 16 July 1872 vol. 212 c1276/PP.

agricultural labourers—the class which was in the worst condition, and which had no traditions of self-help' had managed to 'rise in their own defence against their employers, and show themselves capable of helping themselves'. He feared that if the Parliament would 'let any class of the people be accustomed to believe that the House of Commons might be made a machine for their special interest,' then 'it would be no longer possible to look forward to a state of contentment between classes'.[53] Herbert was not alone in this opinion. In light of a long tradition of voluntary arrangements between workers and employers, other members of Parliament 'could not see why that House should enter on a system of legislation which would be an interference with the labour question throughout the whole country'.[54] Even supporters of working time reduction were concerned about the 'danger of over-legislation',[55] but believed that the case of holidays was exceptional. MP Anthony J. Mundella from the Liberal Party described the bank holidays as a 'legitimate interference with freedom of contract', and accused Herbert for merely 'protecting the pockets of the rich'.[56] MP Joseph Cowen 'believed that if fixed holidays were agreed upon throughout the country fewer irregular holidays would be taken', thus benefiting both workers and manufacturers.[57]

Having lost the war on legal intervention in cases of adult workers, the anti-legislation camp turned to other argumentations, sometimes contradictory. While some claimed that legislative intervention is redundant since workers have secured de facto the same conditions, others argued that these conditions are in fact harmful and that the workers themselves do not want them. MP James Stevenson, for example, explained that although 'without any legislation whatever on the subject', holidays are 'actually taking place . . . by the voluntary consent of the parties interested', and there is in fact 'no desire on the part of the working classes for a measure of this kind, who would lose their day's pay by the enforced holiday'.[58] MP Charles Norwood raised a similar concern saying that enforcement of holidays would harm the poorest workers as 'the wages that might have been earned on those days would be a serious loss to them'.[59] Norwood repeated this argument a few months later, stating that: 'The working classes, unfortunately, had too many holidays in the course of the year, it was very hard upon the men working at the docks, railways, and shipping, that for an additional four days in the year they should be precluded from earning their living.'[60]

The argument that workers did not want holidays clearly contradicted evidence of workers avoiding overtime as much as possible and taking holidays against their employers' will as well as the fact that for decades workers have been

[53] Ibid. [54] HC Deb 6 April 1875 vol. 223 c393/PP.
[55] HC Deb 16 July 1872 vol. 212 c1277/PP. [56] Ibid.
[57] HC Deb 6 April 1875 vol. 223 c396/PP. [58] HC Deb 24 February 1875 vol. 222 c802/PP.
[59] Ibid. c803. [60] HC Deb 6 April 1875 vol. 223 c394/PP.

negotiating their free time with their employers.[61] Nevertheless, this contradiction was actually quite real in the workers' experience. For many workers, holidays, and certainly unpaid holidays, had to be considered within a wider balance between time and money. If they had the choice, they would often choose higher wages over longer periods off work.[62] Some feared that 'after a certain limit, compulsory holidays are to the detriment of the employee rather than to his advantage', as it might have meant 'probable loss of pay, and possibly out-of-pocket holiday expenses in addition'.[63] Following interviews with workers, a 1912 parliamentary report on the Shops Act thus noted that although '[l]ike everyone else, these men desire as much holiday as possible, other things being equal', in reality 'many of them feel that excessive holidays would reduce their already low wage beyond living rate'.[64] It is therefore understandable why MP George Lansbury from the Labour party warned that a system of compulsory unpaid holidays 'really means enforced semi-starvation for the women and children of millions of workers in the country', and criticized his fellow colleagues 'that, while we shall be going away for a well-earned holiday and those on the Treasury Bench will not have their wages stopped... the workers and the people who find the money for our holidays will receive no wages for these enforced holidays'.[65]

For some of those opposed to holiday legislation, the problem with workers' holidays was not the question of legal regulation, but the provision of holidays itself. Arguing against the 1875 extension of the Bank Holidays Act MP George Leeman from the Liberal Party claimed that even without legislation workers are 'able to take a holiday when they pleased', and that consequently they 'had run holiday mad'. He thus 'feared the tendency of the Bill would be to increase the evil'.[66] Others feared that more holidays would increase the alcohol consumption of the working classes.[67] The notion of the worker as a 'holiday mad' drunk was common. For a parliamentary report Rev. J. Fraser explained that 'the English working man is far behind the continental working man in his conception of the mode in which it beseems a rational creature to spend a holiday. His notion of pleasure is too often coarse and animal; the idea of beer is the predominating one in his mind.'[68] Holidaymaking was

[61] See for instance HC Deb 24 February 1875 vol. 222 c800/PP; A. F. Stanley Kent, 'Second Interim Report on an Investigation of Industrial Fatigue by Physiological Methods, by A. F. Stanley Kent, M.A., D.Sc. (Oxon), Henry Overton Wills Professor of Physiology in the University of Bristol' (Cd. 8335, 1916), 22/PP.

[62] Cross, 'Time, Money, and Labor History', 6.

[63] Reports to the Secretary of State for the Home Department on the Shops Act (Cd. 6579, 1912), 7/PP; see also Pimlott, *The Englishman's Holiday*, 151–3.

[64] 'Reports to the Secretary of State for the Home Department on the Shops Act' (Cd. 6579, 1912), 7/PP.

[65] HC Deb 12 April 1911 vol. 24 c497 / PP.

[66] HC Deb 24 February 1875 vol. 222 c801 / PP.

[67] Ibid. c800.

[68] Commission on the Employment of Children, Young Persons, and Women in Agriculture (1867), First report of The Commissioners, with Appendix Part I. to the First Report. Reports from the Assistant Commissioners and Miscellaneous Evidence (4068, 4068-I, 1868), 44 / PP.

something to be learned, and educating people to make holidays in a civilized way was framed as a civil moral duty. Legislation, Fraser expounded, would not help: 'the law cannot step in to mend things here. It is a moral evil, to be cured only by moral remedies. It needs for its cure a greater sense of responsibility in the masters, as well as a greater strength of principle in the men.'[69]

The bourgeois prejudice about the working classes and their leisure habits was, indeed, a major obstacle in the road to holiday legislation. Many members of the middle classes simply did not like to think that 'everybody should make holiday altogether'.[70] The anonymous contributor to *Vanity Fair* 'passes his miserable four days cursing Sir John Lubbock, and the collective wisdom of Parliament, which in attempt to make all men happy by holidays has only made the greater number idle and drunken, and the rest miserable and helpless'. In his eyes, the legislation led to 'constantly recurring holidays which scarcely leave a man time to get sober before he is bound by Act of Parliament to get drunk again'. Therefore, he continued, 'if it is absolutely necessary that everybody should idle, riot, and get drunk simultaneously ... we should know what to expect and when to expect it, and should make our preparations to go abroad accordingly'.[71] The writer stressed that those who oppose the holiday legislation did not do it because they denied the universal right of holidaymaking. Their problem, he explained,

> is not the holiday itself, but the generality of it, the holiday by Act of Parliament which forces all men to join in it at one and the same time, instead of each man taking it at a different time. And they point to the effect of the general holiday as amply justifying their objections ... It is probably inevitable that holidays should come. They come badly enough every Saturday afternoon; they come also every Sunday and they come at Easter, at Whitsuntide, and at Christmas, besides other times. Now each set of these holidays involves not only the holiday itself, but also a preceding and succeeding stage of idleness, riot and drunkenness, of days when shops are closed, when trains don't arrive, when letters are delayed, and cheques can't be cashed.[72]

Arguments against the holiday bill and its later extensions usually avoided direct statements opposing the right for holidays. MP Benjamin Rodwell admitted that 'it might seem invidious to oppose a Bill which provided for the recreation of the people', but went on to explain that the bill 'would interfere with private enterprize [sic]', and would therefore be 'introducing a new and very dangerous principle'.[73] Indeed, in addition to bourgeois reservations and the argument that legislation

[69] Ibid. 44–5 / PP.

[70] Anon., 'Holidays', *The Falkirk Herald and Linlithgow Journal (published first in Vanity Fair)*, 9 May 1878, 6 / BNA.

[71] Ibid. [72] Ibid. [73] HC Deb 24 February 1875 vol. 222 c806 / PP.

was overkill, the main opposition to holiday legal regulation—in parliament and outside of it—focused on its potentially negative economic effects.[74] Many believed that holiday legislation would mean that 'the whole work-day life of the country is forcibly stopped',[75] resulting in serious harm to the economy.[76] Moreover, some feared that giving British workers longer holidays than other nations give their workers will lead to an industrial disadvantage on the international level.[77] Extending the Act to non-clerical workers raised the fear that the banks and the post offices would cease to function optimally and that railway and hackney carriage traffic would suffer delays or might even stop.[78] Clearly, the objection to workers' holidays reflected the worries of the upper classes regarding their possible effects on their routine life and business.

Justifying workers' holidays

How then were workers' holidays justified? I argue that health, although overlooked in previous studies, played a key role in proponents' arguments. Once experts demonstrated that work could lead to pathological conditions, health acquired legitimacy as a premise in the positions of those supporting legal interference in the conditions of adults' labour. In the following, I will present three ways in which health concerns influenced the arguments in favour of reduction of working time and later on holiday legislation: the precedent of labour laws for women, a new ideology of productivity, and the inclusive character of the emerging emotional pathologization of work.

The paradoxical outcome of labour legislation for women

As mentioned earlier, reduction of working time was generally accepted for children and women starting from the 1840s. The common argument was that women and young people were weaker and generally less healthy than men. Nonetheless, labour legislation for women was a crucial precedent that facilitated its extension to men. As MP Herbert Gladstone noted:

[74] Barton, *Working-Class Organisations and Popular Tourism*, 89.

[75] Anon., 'Holidays', *The Falkirk Herald and Linlithgow Journal (published first in Vanity Fair)*, 9 May 1878, 6 / BNA.

[76] HC Deb 6 April 1875 vol. 223 cc396–397 / PP.

[77] HC Deb 24 February 1875 vol. 222 c800 / PP.

[78] Ibid. Anon., 'Holidays', *The Falkirk Herald and Linlithgow Journal (published first in Vanity Fair)*, 9 May 1878, 6 / BNA.

> You cannot say that that measure [regulation of shop hours] does not interfere with adult labour simply because it limited the hours of women and children. That argument is a mere quibble. Everybody knows that it was a direct interference with adult labour, and has been so down to this day... the bulk of the Conservative party have again and again consented, practically and directly, to interfere with adult labour, and a great many of them have voted for this Bill. There is a reason for the interference with adult labour in this case.[79]

Beginning in the late 1880s, support for applying working hour regulations to all workers 'without regard to the age or sex of the persons employed' was becoming increasingly common.[80] In the report from the Select Committee on the Shop Hours Bill of 1892, which aimed to limit the employment of women to 74 hours per week, almost every interviewee was either questioned specifically about the question of gender equality or mentioned it voluntarily, as in the following dialogue:

> – I think legislation is very good for women and for men as well...
> – You would even include men as well?
> – Oh, yes, certainly.[81]

As we have seen in the debate on legal interference, a decade earlier, claims that state regulation should be applied to both sexes would have been seen as standing outside the norm.[82] Indeed, this position was not yet accepted by most legislators, and in some cases, it even led to confrontations, as in the committee's interrogation of Thomas Sutherst, the author of the book *Death and Disease Behind the Counter* (1886):

> – ...you do not answer the question I put except by stating that you want the Bill in effect to go much further than it does; you want all men to be protected?
> – Yes.
> – I have not asked you about men at all... Although this Bill does not go so far as you and others desire it to go, you have no hostile feelings to this Bill on that account?
> – Not at all.
> – You would like this Bill to become law?

[79] Miners' Federation of Great Britain, Coal Mines Eight Hours Bill: Transcript of the House of Commons Second Reading, 22 June (Manchester 1908), 9 / TUC/HD 9551.5.
[80] Select Committee on the Shop Hours Bill (HC 287, 1892), x / PP. [81] Ibid. 3.
[82] Ibid. 3, 95.

- Assuming that I thought that women would not suffer, or that the number of women employed would not be diminished by this Bill passing, I am in favour of it, certainly.[83]

By the early 1890s, there was a growing demand that equal legislation be passed. Sutherst's answers hint at the central concern that informed this demand: the fear that women would suffer even more if the Bill were to become law. Trade unionists, members of the early closing associations, and others worried that regulation of working hours and longer holidays only for women would result in male workers being more attractive to employers and women's wages being decreased, thereby worsening their already difficult employment conditions.[84] Thus, referring to the Shop Hours Bill, the Women's Liberal Federation resolved

> [t]hat any benefits which are likely to be conferred upon women and young people . . . should be extended to men, because, in the opinion of this council, no distinction should be made between the sexes which would place any impediment in the way of the employment of women, or which would have a tendency to make women's labour cheaper than male labour.[85]

The well-known social reformer Laura Ormiston Chant, then member of the Women's Liberal Federation, requested that she be allowed to read this resolution in front of the Select Committee on the Shop Hours Bill; she added that it was 'voted upon by 730 women, delegates from all over England, part of Scotland, and all Wales, and with very few dissentients'.[86] Ormiston Chant's answer to the committee's question as to whether she had 'any personal views in the matter' is worth quoting in full:

- Yes, I have very strong ones.
- Will you state what they are?
- May I say, to begin with, that my ideas are very strong that such a Bill is wanted, that the long hours are most disastrous to the young women employed in close quarters.
- Have you given any special study to the employment of women of this particular kind?
- Yes, I have. I was a sister in the London Hospital for about two and a-half years and we had a great many cases in of girls whose health was quite wrecked by their long hours . . . I myself am exceedingly concerned in finding employment for numbers of women, and the difficulties are so very great, and the pay so small, that anything that would lessen the pay and increase

[83] Ibid. 59. [84] Ibid. 63, 293. [85] Ibid. 203. [86] Ibid.

the difficulty, would be adding to the terrible nature of the work one is having to do in another way. I have done 27 years of rescue work.
- Therefore you would prefer a Bill which would apply to both sexes equally?
- Yes.[87]

Three years later, the Trade Union Congress adopted a similar resolution with a wider scope, stressing that 'in all legislative enactments relating to labour, so far as it affects hours of work and remuneration, there should be no distinction made as to the sex of the person so employed, and that perfect equality of treatment should be adopted'.[88]

Yet some were convinced that legislation for women would not have any negative effect, either because women's wages were already dramatically lower than men's, and thus that employers would continue to prefer them even if limits were placed on their working hours, or because in many occupations, a clear gendered division of labour was already institutionalized in practice.[89] This was also Sidney Webb's opinion, who supported equal regulation but believed that the unions should accept legislation for women 'even if Parliament persists in denying it to the men'.[90] Webb explained:

> It is frequently asserted as self-evident that any special limitation of women's labor must militate against their employment. If employers are not allowed to make their women work overtime, or during the night, they will, it is said, inevitably prefer to have a man. Thus, it is urged, any extension of Factory legislation to trades at present unregulated must diminish the demand for women's labor. But this conclusion, which seems so obvious, really rests on a series of assumptions which are not borne out by facts . . . To begin with, in over nine-tens of the industrial field there is no such thing as competition between men and women: the men do one thing, and the women do another. There is no more chance of our having our houses built by women than of our getting our floors scrubbed by men.[91]

Paradoxically, the demand for equal legislation for men and women emerged as a result of the political weakness of female workers together with the presumption that women were weaker than men, particularly in regards to the potential physical effects of labour.

[87] Ibid.
[88] Report of the Trades Union Congress, Cardiff, 2 September 1895, 58 / TUC/HD 9551.5.
[89] Select Committee on the Shop Hours Bill (HC 287, 1892), 3, 66 / PP.
[90] Webb, *Women and the Factory Acts* (1896), 4. [91] Ibid. 10–11.

Medical knowledge and economic efficiency

A second way in which health considerations influenced the question of working time was through the medicalization of research on efficiency, which legitimized rest as necessary for optimal production outputs, thus saving it from being condemned as laziness. The reduction of working hours has puzzled many labour historians and historians of late Victorian Britain. Manfred Bienefeld and Gary Cross take different approaches to addressing the question: Bienefeld examines the effect of macro-economic conditions on the 'mechanics of bargaining', and Cross analyses the interest-matrix between workers/employers and money/time.[92] Still, both of them note that the viability of decreasing working hours was related to a transformation in the concept of 'productivity'. Cross, more specifically, refers to a new 'understanding of productivity', and new 'productivist optimism', which 'linked efficiency with improved working conditions and especially shorter working hours'.[93] However, both studies remain unclear on what exactly led to the change in the productivist ideology.

Two points should be made clear when evaluating the real efficacy of shifting conceptions and evaluations of productivity. First, utilitarian arguments on the significance of workers' holidays for the industry and economy on the one hand, and claims regarding the benefits of leisure for workers' health, on the other, were frequently intertwined, as workers' health was seen as a precondition for efficiency. This was a characteristic of the late nineteenth-century discourse on work in Britain as well as in the Unites States, France, and Germany, as demonstrated thoroughly in Anson Rabinbach's seminal book *The Human Motor*.[94] Indeed, in the debate on working hours and holidays, the comparison between workers and machinery appeared frequently in the British press throughout the period. In 1872, an anonymous author explained that

> if a man as a mere instrument of labour is to last for forty years, it will be found that eight hours of energetic work given out day by day, year after year, is as much as the machine can stand without deterioration; and moreover, the best experienced [sic] has shown that the skilled man who works with a will can turn out as much work in eight hours as when he is compelled to spend ten hours over the same task. Eight hours of skilled labour should therefore constitute a legal day's work.[95]

[92] Bienefeld, *Working Hours in British Industry*; Cross, 'Time, Money, and Labor History'.

[93] Cross, 'Time, Money, and Labor History', 9.

[94] Rabinbach, *The Human Motor*; for the German context see also the section 'The Problem of Emotions in Industrial Society' in Donauer, *Emotions at Work—Working on Emotions*.

[95] Anon., 'The Peers and the Working Classes', *The Derby Mercury*, 10 January 1872, 8 / BNA.

In 1904, a commentator advocated for bank holidays by arguing that while 'the working parts of a machine can be renewed, the nerves and physical powers of a man once destroyed cannot be made good. Consequently he must be given a rest, he must work shorter hours.'[96]

In response to opponents' scepticism concerning its potential economic implications, supporters of holiday legislation followed the efficiency argumentation of their political ancestors The Eight Hour Day Movement and The Early Closing Association, namely that 'the highest efficiency is best attained by sacrificing a reasonable share of time to rest and play'.[97] It was also stressed that most employers agreed 'that shorter hours and good conditions, together with the maintenance of a fair standard of wages, do lead to the increased efficiency of the workman'.[98] Many employers admitted that they would close early and provide more holidays if the rest of the employers would do the same. A 1910 survey on early closing found out that '[o]ut of 140 who were asked their views about compulsory early closing, 83 were definitely in favour of it, while only 35 objected to it, the remaining 22 declining to commit themselves'.[99] Trade unionists pushing for the legislation thus left no rhetorical stone unturned, proclaiming that the employers themselves were 'the victims of a system which they themselves deplore, but from which they have no way of escape'.[100] When regulation had put an end to the escalating competition, proponents said, employers would gladly improve worker's conditions.

A second point that should be raised when considering the efficiency argument is the role of medical knowledge in this debate. As Rabinbach shows, and as Arthur McIvor analyses in detail in the British case, the development of industrial medicine in the late nineteenth century had a major impact on the emerging consideration of 'the human element' in production processes and on the new linkage between individual health and economic efficiency.[101] Scientists and reformers were attempting to replace the moral emphasis that characterized most of the nineteenth-century European discourse on work with what was understood as rational and objective claims. And indeed, the study of fatigue and mental capacities, including numerous investigations of the correlation between overtime and output, converted medicine into the expert authority on

[96] Anon., 'Bank Holidays', *The Western Times*, 5 April 1904, 7 / BNA.

[97] Anon., 'The Development of Holidays', *The Bristol Mercury*, 27 December 1895, 6 / BNA.

[98] Crosfield, A Plea for the Eight-Hours' Day in Continuous Processes (1912), 5 / TUC/HD 5106.

[99] Margaret H. Irwin, *Exempted Shops: Report I on Tobacconists, Newsagents, Fruiters and Confectioners or Sweet Shops, Which Are Exempted from the Early Closing Order under the Present Shop Hour Act in Glasgow* (Glasgow: The Scottish Council for Women's Trades, 1910), 4 / TUC HD 5106.

[100] Ibid. 4; see also Select Committee on the Shop Hours Bill (HC 287, 1892), 3, 47, 60 / PP.

[101] Rabinbach, *The Human Motor*, 9–11; McIvor, 'Employers, the Government, and Industrial Fatigue', 724.

issues of work efficiency.[102] Medical doctors were the leading voices challenging the common assumption that the 'simplest method of increasing output... is to increase the number of hours in the working day'.[103] Physicians were now arguing the opposite. Basing their reports on scientific experiments, they made points that gave scientific weight to the movement to reduce working hours, such as that '[t]wo or more overtime days produce a proportionately greater lowering of efficiency', or, more generally, that overwork is 'physiologically and economically extravagant' and it 'frequently fails in achieving its object'.[104]

As one might expect, there was a gap between the new body of theory and actual management practices. Nonetheless, by the First World War, the claim that output is actually higher when workers work less hours and are provided with intervals for rest was being echoed by politicians and trade unionists alike.[105] While such arguments can be found in the medical literature going all the way back to the 1870s, it was the war emergency that led the government to support and sponsor the extensive investigation of 'industrial fatigue'.[106] The Health of Munitions Workers Committee resolved in 1915 that 'provision for workers' health, welfare, and the elimination of industrial fatigue was not only desirable socially, on humanitarian grounds, but essential for the maintenance of industrial efficiency'.[107] This opinion was gradually embraced by the majority of British employers.[108]

The new hegemonic position of medical expertise in the study of economic efficiency helped to solve the ongoing debate over working time, health, and productivity, strongly favouring holiday legislation. While traditional moralistic arguments regarding workers' behaviour and efficiency emphasized idleness as the source of unproductivity, from the 1870s onwards, these positions were reconceptualized through the research on fatigue and neurasthenia as the only obstacle to perfect economic growth. As a result, regulated periods of idleness were de facto approved as being part of the optimal work process.

[102] Rabinbach, *The Human Motor*, 5–6; McIvor, 'Employers, the Government, and Industrial Fatigue'.

[103] Second Interim report on an investigation of industrial fatigue (Cd. 8335, 1916), 7 / PP; see also Arthur Shadwell, *Industrial Efficiency: A Comparative Study of Industrial Life in England, Germany and America* (London: Longmans, Green and Co., 1906); McIvor, 'Employers, the Government, and Industrial Fatigue', 725, 730–1.

[104] Interim report on an investigation of industrial fatigue by physiological methods (Cd. 8056, 1915), 3 / PP; Second Interim report on an investigation of industrial fatigue (Cd. 8335, 1916), 7 / PP; Second Interim report on an investigation of industrial fatigue (Cd. 8335, 1916), 22 / PP.

[105] Walls, 'Conditions in British Iron and Steel Works' (1912), 4 / TUC/HD 5106; Crosfield, 'A Plea for the Eight-Hours' Day in Continuous Processes' (1912), 5 / TUC/HD 5106; Anon., 'The Peers and The Working Classes', *The Derby Mercury*, 10 January 1872, 8 / BNA.

[106] Interim report on an investigation of industrial fatigue (Cd. 8056, 1915), 3; Second Interim report on an investigation of industrial fatigue (Cd. 8335, 1916), 22 / PP; McIvor, 'Employers, the Government, and Industrial Fatigue', 731.

[107] Quoted in: McIvor, 'Employers, the Government, and Industrial Fatigue', 730–1.

[108] Ibid. 729–30.

Parliament and the pathologized worker

Whereas the previous points primarily help explain the historical reasons at the foundation of the process of reducing working time, holiday legislation exemplifies a third effect of rising concerns about workers' health: namely, the emotional pathologization of the worker. Medical knowledge was not only instrumentalized for the research of efficiency—it also had crucial consequences for the legal protection of citizens' health. By the beginning of the twentieth century, the argument against Parliamentary intervention in the affairs of private business was opened to an exception, and regulation came to be accepted by more MPs and commentators if 'in the first place you interfere for the benefit of the health and safety of those who work in it'.[109] This was a new thought. Indeed, health considerations were present in the debates on working time, but the focus there was to prevent injuries and accidents by keeping workers more alert. As Susan Barton rightly notes, in these cases public opinion played a central role: for example, rail worker's fatigue posed a potential danger to the general public.[110] In contrast, in the context of holiday legislation, the argumentation focused on nervous and emotional problems that arose as a 'spontaneous' condition (and not as a result of an accident) and that did not necessarily endanger people other than the workers themselves.[111]

The medical knowledge on the nervous and emotional consequences of work was known to the key players in and around the legislature and had a serious influence on the debates on holiday rights. In Parliament, this influence manifested itself first and foremost in MPs' deep acquaintance with the medical literature and their vast knowledge of conditions such as fatigue, neurasthenia, and overwork.[112] Starting at the turn of the century, Parliament also appealed for professional reports on these topics, a trend that gained intensity during the war.[113] Thus, for example, the detailed 1915 report on Industrial Fatigue began with the following declaration: 'The starting point for the present inquiry may

[109] HC Deb 31 March 1911 vol. 23 cc1739–1740 / PP.

[110] Barton, *Working-Class Organisations and Popular Tourism*, 115; see also George W. Alcock, *Fifty Years of Railway Trade-Unionism* (London: Co-operative Printing Society Limited, 1922), 247–8; Walls, 'Conditions in British Iron and Steel Works' (1912), 2 / TUC/HD 5106.

[111] Departmental Committee on Compensation for Industrial Diseases (Cd. 3496, 1907), 280 / PP.

[112] For instance HC Deb 15 April 1890 vol. 343 c588 / PP; HC Deb 19 June 1891 vol. 354 cc923 / PP; HC Deb 21 March 1893 vol. 10 c733 / PP; HC Deb 14 November 1893 vol. 18 c951 / PP; Departmental Committee on Compensation for Industrial Diseases. Report of the Departmental Committee on Compensation for Industrial Diseases. Report (Cd. 3496, 1907) / PP; HC Deb 08 May 1911 vol. 25 c1002 / PP; HC Deb 19 July 1911 vol. 28 c1161 / PP; McIvor, 'Employers, the Government, and Industrial Fatigue', 730.

[113] HC Deb 21 March 1893 vol. 10 c732 / PP; Departmental Committee on Compensation for Industrial Diseases (Cd. 3496, 1907), 261, 280 / PP; HC Deb 23 September 1915 vol. 74 c574 / PP; Interim report on an investigation of industrial fatigue (Cd. 8056, 1915), 3; Second Interim report on an investigation of industrial fatigue (Cd. 8335, 1916), 7 / PP; McIvor, 'Employers, the Government, and Industrial Fatigue', 724–32.

perhaps be dated from a resolution adopted at the International Congress on Hygiene and Demography held at Brussels in 1903 that "the various governments should facilitate as far as possible investigation into the subject of industrial fatigue"'.[114]

The questions addressed to interviewees in Parliamentary committees also reflect the MPs' interest and knowledge in the medical consequences of overwork. Based on the claim that '[o]ne of the principal reasons why such legislation as the interference with the liberty of the subject is brought forward is because of the question of health',[115] committees summoned physicians and other experts to testify and questioned ordinary workers about their own experiences.[116] Remarkably, both MPs and interviewees who were not themselves physicians cited professional opinion to strengthen their arguments.[117] A telling example can be found again in the words of Ormiston Chant. Following her call for equal legislation, Ormiston Chant added: 'might I be allowed to add that my husband is a doctor, and he has a very large practice, amongst poor people generally, working people, and his experience is that the Bill is as much needed for men as it is for women'.[118] A similar strategy is pursued in trade union documents. Unions appointed consulting physicians who represented the workers in front of Parliamentary committees, and the unions literature's repeated claim that overtime is injurious to health was backed up using professional studies, for example from *The Lancet*.[119]

As we have seen, the medical discourse on overwork did not provide clear definitions of its causes and its symptoms. This was no different when it was taken up and applied in political discourse. Most of the time, Parliamentary discussions on the effects of overwork proceeded in very broad terms, usually referring generically to the 'excessive burden of overwork' or workers' 'health' without further elaboration.[120] The Parliamentary report on industrial fatigue contained a preliminary remark explaining that fatigue was used in the document 'in connection with various structures and baring different meanings accordingly' and that it would thus serve in the report 'as a general term, and may be taken to include a general lowering of the functions of the body as a result of extreme or

[114] Interim report on an investigation of industrial fatigue (Cd. 8056, 1915), 3.

[115] Select Committee on the Shop Hours Bill (HC 287, 1892), 63, 7 / PP.

[116] Ibid. 67, 105; HC Deb 21 March 1893 vol. 10 c733 / PP; Stearns, *Lives of Labor*, 249; Hosgood, 'Shops, Shop Assistants, and Shop Life', 326–7; McIvor, 'Employers, the Government, and Industrial Fatigue', 724–5.

[117] For instance Select Committee on the Shop Hours Bill (HC 287, 1892), 53, 64, 293 / PP.

[118] Ibid. 293.

[119] Tuckwell, *Women's Work and Factory Legislation* (1895), 8; Report of the Special Commission of Hours of Labour in Continuous Industries (1912), 19 / TUC/HD 5106; McIvor, 'Employers, the Government, and Industrial Fatigue', 730.

[120] HC Deb 2 May 1888 vol. 325 cc1001–1002 / PP; Select Committee on the Shop Hours Bill (HC 287, 1892), 7, 53, 105 293 / PP; HC Deb 19 June 1891 vol. 354 cc921–2 / PP; HC Deb 31 March 1911 vol. 23 c1741 / PP.

prolonged activity'. The report also mentioned that fatigue might be aggravated by '[a]ny interference with the normal functions of the body' or by conditions which are 'exceedingly variable and their scientific study presents exceptional difficulties'.[121] As if this were not vague enough, the report continues stating that 'from the industrial point of view, it is less important to understand the deep-lying causes than to be able to recognise and control the circumstances which have led up to them in the past and are likely to lead up to them in the future'.[122]

The significant point to make is that through the focus on emotional and nervous burdens, the 'health' argument did more than facilitate the passage of legislation; rather, it enabled an increasingly egalitarian legislation. In the debates on physical disorders caused by work, it was generally accepted that women and children were much more affected than men, or, in other words, that 'men have always been regarded, and we as men know, or assume at any rate that we are healthier than women are'.[123] But when it came to the nervous and emotional maladies of work, the argument was applied equally to both sexes and helped to legitimize the legal interference in the conditions of adult (including male) labour. The Parliament discussed cases of male workers who, 'in consequence of overwork', became 'temporarily incapacitated from duty owing to mental derangement', or who 'have been driven to commit suicide by overwork'.[124] MPs were referring to both men and women when they complained, for example, that both '[m]asters and mistresses of schools... were so over-worked that they often became quite irritable, and lost their temper with the children', thus converting school from 'a place of happiness and contentment' to 'a place of torture'.[125] More generally, Parliamentarians sympathetic to the plight of workers argued that 'it is clear that in a large number of cases men and women are employed 80, 84, and 90 hours a week in occupations which, if not physically exhausting in the sense that manual labour is, yet are destructive, when prolonged, of physical, mental, and even of moral health'.[126]

The effect that the nervous and emotional pathologies caused by overwork had on all classes and occupations was likewise echoed in the legislative debate. Discussing the working conditions of telegraph clerks, MP Earl Compton reported on the high percentage of clerks retiring due to 'physical and nervous affections... as well as mental fatigue'.[127] In 1899, considering the suggestion that retailers be mandated to supply chairs for shop assistants, the Lord Bishop of Ripon referred to the outcomes of fatigue and nervous exhaustion, stating that:

[121] Interim report on an investigation of industrial fatigue (Cd. 8056, 1915), 4. [122] Ibid.
[123] Select Committee on the Shop Hours Bill (HC 287, 1892), 121, see also 47 / PP.
[124] HC Deb 20 June 1881 vol. 262 cc836 / PP; HC Deb 4 January 1894 vol. 20 cc813 / PP.
[125] HC Deb 26 July 1883 vol. 282 cc595 / PP. [126] HC Deb 21 March 1893 vol. 10 c741 / PP.
[127] HC Deb 15 April 1890 vol. 343 cc588 / PP.

> The preservation of health, in the end, is a great economy... Customers leave a shop because of what they call the impertinence or the indifference of the person who is serving them. Is not the apparent bad humour on the part of the shop-assistant largely due to her having been standing for so long a time? When we complain of the ill-temper of those who serve us, we often forget that ill-temper is the result of fatigue mid nervous exhaustion.[128]

That every worker had to be protected—miners, weavers, shop assistants, businessmen, and even students—increasingly became a matter of legislative common sense. In 1890, the House of Commons discussed a report from 1886 that referred to the retirement of clerks due to 'physical and nervous affections... as well as mental fatigue'.[129] In 1911, MP James Dawes demanded to know how many workers in the London postal service had died in the past five years 'as the result of mental and nervous diseases'.[130] The report on industrial fatigue of 1916 noted explicitly that '[m]embers of the office staff may show signs of fatigue and overstrain equally with the factory hands'.[131] Based on similar reports produced during the war, MPs contended that signs of fatigue and overwork were popping up in diverse occupations and were affecting manual labourers just as well as white-collar workers.[132]

Putting this knowledge into action for legislative objectives was clearly very difficult, and it seems that it was equally hard to define an optimal remedy. Added to the fact that the pathology afflicted both genders and all classes, it was also said to be invisible, meaning that even if a worker looked healthy, dangerous nervous and mental conditions could develop rapidly without early signs.[133] As physicians warned that the effects of overwork accumulate and are not limited to a specific day of work, and since the primary medical advice was to rest and avoid work, holidays for all became the logical legislative response to the issue.[134] This was indeed the result. After the second decade of the twentieth century, new holiday bills were addressed to all 'workpeople', securing each year 'at least seven whole holidays'.[135] In addition, more and more measures slowly succeeded in securing one weekly day of rest and in creating the weekend.[136] Needless to say, it took time

[128] HL Deb 11 July 1899 vol. 74 cc457 / PP. [129] HC Deb 15 April 1890 vol. 343 c588 / PP.

[130] HC Deb 08 May 1911 vol. 25 c1002 / PP.

[131] Second Interim report on an investigation of industrial fatigue (Cd. 8335, 1916), 22 / PP.

[132] HL Deb 25 February 1919 vol. 33 cc302 / PP.

[133] Burford and Hey, *Neurasthenia in Women* (1907), 13; Dowse, *On Brain and Nerve Exhaustion* (1880), 24.

[134] See, for instance, the Second Interim report on an investigation of industrial fatigue (Cd. 8335, 1916), 7, 22 / PP.

[135] Holidays. A bill to secure one day's rest in seven for all workpeople; to make the first day of May a Bank Holiday throughout the United Kingdom to secure a week's holiday for workpeople; and to secure payment of wages for public holidays (Bill 242, 1911), 1–2 / PP.

[136] Weekly rest-day. A bill to secure one day's rest in seven for all workpeople; to make the first day of May a bank holiday throughout the United Kingdom; to secure a week's holiday for workpeople; and

for this wave of holiday bills to become actual acts and to be consistently implemented. The process was ultimately completed with the passage of the Holidays with Pay Act of 1938. Indeed, in 1938, the Lord Amulree, William Warrender Mackenzie, as the Chairman of the Royal Society of Arts and of the Departmental Committee on Holidays with Pay, declared that '[a] holiday cannot be said to be more of a necessity to the one class than to the other'.[137] But there was a long way to go before this position took on the irrefutability of common sense.

In parallel to the holiday legislation, bargaining and other methods did not cease to play a role. In fact, as Peter Stearns shows, the real impact of collective bargaining doubled in the years before the war.[138] In some cases, collective arrangements between employers and employees prolonged the legal annual holiday, or helped to secure an eight hour standard working day, and the unions continued their negotiations with the Employers' Association to improve working conditions without having to resort to strikes.[139] Nonetheless, the British holiday legislation of the early twentieth century was path-breaking.

Hard work, seaside, and happiness

The political debate on workers' holidays thus took a complicated course that ultimately climaxed in the passage of holiday legislation for all workers in 1938. But why was the seaside considered to be the ideal destination for spending these new blocks of free time? Barton notes that '[t]ime spent at the seaside in comparative idleness had to be justified on health grounds',[140] and other historians claim without further elaboration that health was in fact nothing but an excuse or a mask covering the desire to have fun.[141] However, because none of these historians of the British seaside and workers' holidays examine the interaction between medical knowledge and holidaymaking, they usually end up making broad speculations about the medical and emotional effects of the holiday, sometimes resorting to problematic anachronisms. At the same time, works on the aristocratic epistemology and practices of the water cure usually do not link it

to secure payment of wages for public holidays (Bill 53, 1912), 2; Factory and Workshop Act, 1901, Amendment. A bill to amend the Factory and Workshop Act, by reducing the period of employment in cotton factories to forty-eight hours per week (Bill 199, 1913), 2 / PP; Borsay, *A History of Leisure*, 193, 206.

[137] Amulree, 'Industrial Holidays' (1938), 6. [138] Stearns, *Lives of Labor*, 256.

[139] Barton, *Working-Class Organisations and Popular Tourism*, 79; Walls, 'Conditions in British Iron and Steel Works' (1912), 1 / TUC/HD 5106; Eight Hour Standard Working Day National Defence Found (Draft), 1899 / TUC/HD 5106.

[140] Barton, *Working-Class Organisations and Popular Tourism*, 73.

[141] Borsay, *A History of Leisure*, 91; Beckerson and Walton, 'Marketing the Intangible at British Resorts', 66.

to the political debate on the democratization of holidays.[142] It is here that the difference between efforts to provide opportunities for (urban) rational recreation and the provision of longer holidays should be stressed. As Peter Bailey makes clear, 'rational' urban recreation for workers was mainly a moral project; however, once the concern shifted to workers' health, the solution was shaped by the upper-class imaginary of cure and holidays. Here is where the seaside appears, as watering places were what the British upper classes, including many of the legislators and physicians themselves, had in mind when looking for a cure, while many workers had to learn that holidays, summer, and seaside go together.[143] Indeed, as noted before, some workers, especially in the industrial North West, did visit the seaside and bath in watering places. These practices were not an effect of legislation. The question here is what epistemology of holiday-making is exposed through the legislative process.

More than anything else, it was the health concern that helped to expand the political and legislative question from the reduction of working hours to the provision of consecutive days of holiday. Sports, clubs, music halls, and even urban parks were seen as moral and respectable forms of recreation, but they did not provide 'health'. That was to be found in holidaymaking, which for members of the middle and upper classes meant no less that 'a short eight hours at the seaside'.[144] Half-holidays were not regarded beneficial when it came to health.[145] And even half of Saturday and all of Sunday were not sufficient '[a]fter the fatigue and worry of the week', as 'one poor girl' told Sir John Lubbock. Lubbock told the House of Commons that the girl explained to him: 'I am so thoroughly worn out that my only thought is to rest on a Sunday: but it goes too quickly.'[146] In order to benefit from the strengthening powers of nature, workers needed more than a few hours break. Hence, the health functions of holidaymaking became a central justification for holidays on Saturdays and Mondays. One contributor put the argument in precise terms: a holiday 'from Saturday till Monday gives our antaean workers such renewed strength from their brief repose on the bosom of mother earth, that they leap, so to speak, into the new week with a twofold energy and purpose'.[147] Medical literature went even further, stressing that in cases of

[142] See Weisz, 'Historical Reflections on Medical Travel'; Steward, 'The Affective Life of the Spa'; Adams, *Healing with Water.*

[143] Barton, *Working-Class Organisations and Popular Tourism*, 84–5, 90–1.

[144] Anon., 'Topics of the Week: Holidays', *The Penny Illustrated Paper*, 6 August 1870, 82 / BNA.

[145] Factory and Workshops Acts Commission: Report of the commissioners appointed to inquire into the working of the factory and workshops acts. Appendix C (C.1443, 1876), 13 / PP.

[146] HC Deb 21 March 1893 vol. 10 c733 / PP. The Report of the commissioners appointed to inquire into the working of the factory and workshops acts (1876) makes another argument for prolonging workers' holidays, as it was proved that 'so far as women are concerned', one entire holiday 'appears to be occupied in washing for their families and attending to other necessary household duties' (Factory and Workshops Acts Commission (C.1443, 1876), 113. On the insufficiency of the half-holiday see ibid. 13.

[147] Anon., 'Topics of the Week: Holidays', *The Penny Illustrated Paper*, 6 August 1870, 82 / BNA.

'nervous affections', a healthy and effective stay at the seaside 'should not be for a shorter period than four or five weeks'.[148]

Given that long holidays in watering places for the sake of health were already an established practice for the upper classes—the milieu that produced the majority of legislators—their image of the form and purpose of holidays should be taken into consideration when seeking to understand the extension of the habit to other classes. Thus, for example, the 'class inclusive' medical discourse can be associated with the rhetoric of leisure and 'hard work'. When talking about their own holidays, members of the upper classes frequently contrasted seaside holidays with hard work.[149] 'Work is so hard, so constant, so exacting', claimed an anonymous author as early as 1870, 'that for once in the round of the year we must slacken the girths of labour, if only for a week, and seek some change, even though it may not be perfect rest'. This text reveals that upper-class workers also longed for occasional breaks 'from the healthy claims of toil'.[150] Parliamentary references to the 'hard work' performed by workers from the lower classes is significant because it indicates that MPs recognized workers' toil on a real level. At the same time, it is telling because it indicates the imaginary context in which members of the parliament situated the necessity and function of workers' holidays.

Indeed, 'hard work' is the justification for reduction of working time that appeared most frequently in Parliamentary debates. Already in the second reading of the Bank Holidays Bill in 1871, Lord Robert Cecil, Marquis of Salisbury, said that he did not 'anticipate any opposition to the Bill', given that 'a feeling was generally growing that work in England was quite hard enough, and that additional holidays would not be unwelcome to those to whom they were given, nor unpopular with the general community'.[151] Hard work was mentioned in many cases in which members of the Parliament advocated granting holidays to additional workers and institutions. MP Charles Norwood, aiming to extend the bill to post officers, stressed that 'Post Office employés [sic] were a very hard-worked body of men', and therefore 'fully entitled to the holidays which it was proposed to give to other classes of the community'.[152] In the same meeting, Lubbock claimed that more holidays were justified on the grounds that 'nobody could say that, as Englishmen, we did not work hard enough'.[153]

In addition to their conception of what constituted hard work and its legitimate compensation, it was their own knowledge and imaginary that led politicians,

[148] Aaron Chunder Dutt, *Health Notes for the Seaside: With Special Reference to Whitby and District* (Whitby: Horne & Son, 1895), 19.

[149] On middle class norms and values of in relation to leisure see Mike Huggins, 'More Sinful Pleasures? Leisure, Respectability and the Male Middle Classes in Victorian England', *Journal of Social History* 33, no. 3 (2000): 585–600; Borsay, *A History of Leisure*, 89–91.

[150] Anon., 'Topics of the Week: Holidays', *The Penny Illustrated Paper*, 6 August 1870, 82 / BNA.

[151] HL Deb 4 May 1871 vol. 206 cc129 / PP.

[152] HC Deb 6 April 1875 vol. 223 c396 / PP.

[153] Ibid. c395 / PP.

moralists, and philanthropists to encourage workers to visit watering places. '[T]he sea itself, with its beach, its shrimps, its donkeys, its sand and shell, its piers and jetties' was seen as the ideal destination 'for those who seek a quiet haunt undisturbed by too much importation of the London music-hall and the London "pub"'.[154] The most well-known proponent of the benefits that vacations might have for the working classes was Thomas Cook himself, who provided them with package tours to seaside resorts.[155] But Cook was not alone. In the 1840s, well before the position of workers' holidays had gained widespread traction, Rev. Joseph Brown put into practice his belief that 'bettering the condition of the working classes is a problem affecting the vital interests of society' by taking his parishioners to an annual day at the seaside to enjoy 'healthy and innocent recreation'.[156]

The role of medical knowledge and philanthropists' ideas in the legitimation of holidays is evident in an article from *The Western Times* titled 'Bank Holidays and Health' published in 1901. The author, possibly a physician himself, defended '[t]hose scientific and philanthropic men who ask for more public holidays', stressing that '[t]hey are not faddists, intent only on riding their hobby at any cost'; on the contrary, he explained:

> They know that the social and industrial environment of the masses to-day, the depopulation of the country and the congestion in towns, most speedily affect the vitality if our Race, and the only method of combatting that insidious decay is to tempt people into the open air...[157]

It therefore should not surprise us that in a description of the first bank holidays in May 1871, an anonymous author mentioned 'the many inducements offered by the several railway companies for a "trip to the seaside", and other places of popular resort', which 'were largely taken advantage of'.[158] Likewise, writing in his diary on the second bank holiday, in August 1871, MP John Lubbock noted that '[t]he day was splendid, and the holiday very generally kept. Every seaside place near London, every railway and place of amusement, was chock full'.[159] Certainly, the crowds filling the seaside resorts in the 1870s were mainly members of the middle classes, but going to the resorts soon became a popular habit open to the

[154] Anon., 'Topics of the Week: Holidays', *The Penny Illustrated Paper*, 6 August 1870, 82 / BNA.
[155] Bell and Lyall, *The Accelerated Sublime*, 13.
[156] Anon., 'Working Classes Sea-Side Holiday', *The Morning Chronicle*, 30 September 1857, 3 / BNA. It should be mentioned that at this time one-day excursions into nature were seen as a way to distance the working classes from the bad effects of the city and as part of the Temperance Movement efforts; they were not necessarily related to a belief in the positive effects of nature (Bailey, *Rational Recreation*, 47).
[157] Anon., 'Bank Holidays and Health', *The Western Times*, 28 May 1901, 4 / BNA.
[158] Anon., 'The "Bank Holiday Act" In Operation', *The Daily News*, 30 May 1871, 2 / BNA.
[159] *John Lubbock's Diary 1864–1882*, Supplementary Avebury Papers vol. II / BL / Add. MSS. 62680.

working classes too.[160] Indeed, thanks to the increasingly affordable train and steamboat tickets, more people could 'get far away from the close and smoky town, to the seaside', and thus 'make of every fine Sunday in the summer a holiday indeed'.[161] In 1889, it was '[h]appily . . . an easy thing . . . for even the working-classes to enjoy a holiday thoroughly', as seaside resorts 'which are becoming more and more popular, are as ample as they are cheap'.[162]

The health justification and the established customs of holidaymaking and 'taking the waters' also help explain the institutionalization of summer as the primary holiday season. Holidays in summer were not a workers' custom and were not necessary from workers' point of view, as we can deduce from their habits in periods when the time of holiday was negotiated between employers and employees. In fact, it was more typical for workers to stop their work after Christmas, and indeed, it took a considerable amount of time before the August bank holiday caught on and came to be commonly spent at the seaside.[163] In contrast, the Parliamentary debates suggest that MPs found a summer holiday much more reasonable. The chance to spend large amounts of time outdoors was a major factor when MPs were considering when the bank holidays should take place, and, intuitively, summer offered itself as the best option. Hence, in 1875, Lubbock suggested adding 28 June to the Act, as '[t]here was no Bank holidays between spring and autumn, and he could not help thinking that a summer holiday would be very acceptable to the country'.[164] As an anonymous author stated in 1899:

> the Christmas Bank Holiday can never be an ideal holiday simply because our British weather forbids: the ideal holiday must be spent in the open air. Many of the objections . . . apply to the Easter Monday holiday. It is still too early in the year really to enjoy out-door amusements . . . But amongst legislators and lawyers too there is a feeling that Easter holidays are very unsatisfactory; they are tiresome and unenjoyable; very much on account of the weather. The only ideal Bank Holidays are those of the Summer and the Autumn, and these the masses of the people who derive the greatest benefit from the public holidays enjoy as ideally perfect a manner as their intellectual limitation enable them to do.[165]

[160] Barton, *Working-Class Organisations and Popular Tourism*, 90.

[161] Besant, 'The Amusements of the People' (1884), 344.

[162] Anon., 'The Holidays', *The North-Eastern Daily Gazette*, 10 June 1889 / BNA.

[163] Barton, *Working-Class Organisations and Popular Tourism*, 84, 85, 91.

[164] HC Deb 6 April 1875 vol. 223 cc392–8 / PP. See also HL Deb 16 May 1871 vol. 206 cc861–868 / PP.

[165] Anon., 'Bank Holiday Suggestions', *Saturday Review of Politics, Literature, Science and Art*, 30 December 1899, 822 / BNA.

Numerous articles and essays on work and leisure described the healthy, reviving powers of holidays in nature and specifically at the seaside with strong emotional overtones and flowery language, reasoning that it is in the 'sweet rural scenes' that holidaymakers can enjoy 'that fresh, vivid, and constantly renewing loveliness'.[166] In 1889, one author noted that '[t]he short cessation from toil is a welcome relief, invariably anticipated with pleasure, and looked back upon with feelings of gratitude', and explained that 'rest is beneficial physically and mentally', among other reasons because 'it affords opportunities... for "seeing the world" which brighten our lives, widen our views, and thus have an elevating and stimulating influence'. Criticizing the 'foolish and shortsighted people who grudge the holidays', he then stressed that '[t]hese off days bring health and strength, pleasure and contentment, and all classes of the community could ill afford to have them curtailed'.[167]

An article about the bank holidays published in 1899 declared that '[s]omething more than mere cessation from work is required to make a real holiday'. The bank holidays, claimed the author, were not invented for 'mere recovery from the fatigues of toil', but rather to 'be a real addition to happiness'.[168] In 1901, a writer explained that '[p]ublic holidays are essential to the common weal now', precisely because of the increasing excitement and nervousness of the time.[169] Another author claimed in 1904 that 'the Bank holidays are, properly speaking, an interlude for recuperation, for the repair of wastage, for the recovery of nervous balance', topping off his praise by tracing out the psychological benefit to found in the fact that they 'furnish stores of pleasant memories that last right through long days of dragging toil'. He historicized his argument for the necessity of holidays by reminding readers that '[m]en of an earlier generation had not to contend with the rush and worry that characterise modern conditions of labour', since '[t]he word "hustle" was not found in their dictionaries', and back then 'workpeople had breathing space even if they worked longer hours'.[170]

Around the turn of the century, ideas on the negative effects of work and the need for emotional alleviation were no longer confined to medical circles, but were also abundantly present in political and public discourse. Interestingly, while the emotional terminology came into use in political discourse only after it had gained the status of a commonplace in the medical literature, it was practically absent in the unions' jargon or in the language of the workers themselves. Up until the World War, workers and unions put more emphasis on physical conditions than

[166] Anon., 'Topics of the Week: Holidays', *The Penny Illustrated Paper*, 6 August 1870, 82 / BNA.

[167] Anon., 'The Holidays', *The North-Eastern Daily Gazette*, 10 June 1889 / BNA.

[168] Anon., 'Bank Holiday Suggestions', *Saturday Review of Politics, Literature, Science and Art*, 30 December 1899, 822 / BNA.

[169] Anon., 'Bank Holidays and Health', *The Western Times*, 28 May 1901, 4 / BNA.

[170] Anon., 'Bank Holidays', *The Western Times*, 5 April 1904, 7 / BNA.

on emotional disorders when declaiming the problems of overwork and, in their promotion of holidays, focused more on their moral and physical benefits.[171]

Rising wages and mass transportation clearly facilitated the possibility for wide ranges of people to spend a week or more at a holiday resort, and, by extension, the translation of this possibility into a veritable popular practice. However, they themselves by no means provide a sufficient explanation for the legitimation of workers' holidays, nor for the choice of the seaside as their ideal destination. The present chapter sought to detail the significance of the pathologization of workers in emotional terms and the therapeutic framing of watering places for movements to reduce working time and pass holiday legislation. This process was kicked off with the mid-century Factory Acts, which were based on the assumption that children and women were weaker than men and thus deserved to be secured from the potentially negative physical effects of labour. Beginning in the 1870s, it expanded to include adult men from diverse occupational backgrounds after the argument that they were equally exposed to the dangerous nervous and emotional effects of work gained credence among legislators, doctors, and workers alike. Finally, proponents for universal holidays gained the upper edge when intense research on industrial fatigue during the war subordinated the question of worker efficiency to the recommendations of medical authority, thus solving the riddle of how to balance working hours and productivity.

While urban recreation was regarded as a means for education and moral improvement, nature, and specifically watering places, were increasingly seen by the middle and upper classes as the ideal site for emotional alleviation. By framing the 'problem' of the worker in the language of emotional pathology, medical discourse facilitated the inclusion of the lower classes into the seaside holiday culture. This is not to say that workers did not visit the seaside before the holiday legislation, what I hope to have shown is that holiday legislation gave material expression to a new understanding of emotions and work in the form of concrete rights.

[171] Irwin, *Exempted Shops* (1910); Report of the Special Commission of Hours of Labour in Continuous Industries (1912), 9, 11, 19 / TUC/HD 5106; Walls, 'Conditions in British Iron and Steel Works' (1912) / TUC/HD 5106.

3

Defining the Product

Health, pleasure, and class

It was during the second half of the nineteenth century that the holiday resort was established as a proper commercial industry. What had been little more than small complexes comprised of a bathing facility, a couple of hotels, and basic entertainment (library, coffee house, gardens) developed into a bifurcated system of cooperation between distinct actors, all of whom were pursuing economic profit.[1] This evolution was a specific manifestation of the broader process of economization of Victorian local government.

As a consequence of the Municipal Corporations Reform Act of 1835, which transformed the boroughs into public corporations, and the consecutive Acts that amended it, local authorities expanded their responsibilities and expenditures, particularly by assuming responsibility over health and sanitary services. The expansion of duties together with the new requirement that boroughs make all accounts available for inspection compelled local governments to grapple with issues of financial autonomy and progress. The rising dominance of businessmen in municipal politics during the 1860s and 1870s determined the direction this call for urban financial autonomy and accountability would take, as entrepreneurship, cost effectiveness, and profitability were elevated to key concepts of city government. This process is commonly associated with the so-called 'civic gospel' and the Chamberlain administration of Birmingham. By the end of the century, British local governments were fundamentally tied to liberal bourgeoisie values and ambitions, subjecting the provision of municipal services and other policies to business principles and commercial strategies.[2]

[1] On the development of the British resort from the early eighteenth century onwards, see Borsay, 'Visualising the Seaside, c. 1750–1914'; Peter Borsay and John K. Walton (eds), *Resorts and Ports: European Seaside Towns since 1700* (Bristol: Channel View Publications, 2011). On the influence of the social structure, the processes of land distribution and the type of government on the style of different resorts see Perkin, 'The "Social Tone" of Victorian Seaside Resorts in the North-West'. On the broader debate between an 'urban renaissance' or the emergence of 'leisure towns' in eighteenth-century England, and particularly about the extent of a leisure revolution in the English urban system, see Angus McInnes, 'The Emergence of a Leisure Town: Shrewsbury 1660–1760', *Past & Present* 120 (1988): 53–87; Peter Borsay, 'The Emergence of a Leisure Town: Or an Urban Renaissance?' *Past & Present* 126 (1990): 189–96.

[2] Magnusson, 'Bourgeois Theories of Local Government'; Jones, 'Liberal Businessmen and Municipal Politics in Birmingham, 1865–1900'; Walton, *The English Seaside Resort*, 170. See also Jan Palmowsk, 'Liberalism and Local Government in Late Nineteenth-Century Germany and England',

The emotional economy of holidaymaking: Health, pleasure, and class in Britain, 1870–1918. Yaara Benger Alaluf, Oxford University Press (2021). © Yaara Benger Alaluf. DOI: 10.1093/oso/9780198866152.003.0004

In this context, it is not surprising that the corporation officials of Harrogate, Scarborough, and Blackpool alike, referred to the holiday sector in commercial terms, too, especially since watering places were considered a component of broader health and sanitation services and since most bathing facilities, as well as the sea shore, the gardens, and palaces of entertainment, were owned and operated by the municipal corporations. At first, the municipalities presented their commercial aspirations to the public under a quasi-altruistic ideology. At the opening ceremony of Harrogate's 'Victoria Baths', for instance, Alderman Charles Fortune said that 'the building of that establishment was not purely a commercial question', but that the baths 'were intended for the relief of suffering humanity throughout the United Kingdom, and, indeed, the world'.[3] A decade-and-a-half later, Harrogate's manager, Mr K. J. Buckland, informed readers that the resort was 'treating as many as eleven hundred patients a day' and had therefore 'made £3,300 more than ever before', promising that in the coming year the resort would 'easily surpass that figure'.[4] Indeed, by the beginning of the twentieth century, profit-making was explicitly acknowledged as the corporations' primary goal.[5]

Establishing a commercial industry: advertisers, doctors, and trains

One outcome of the commercialization of local governments was the increased employment of outside experts, such as accountants and advertisers.[6] In principle, the central task of the flourishing British advertising industry was to link the city advertiser to potential consumers in the province and in the colonies, which they

The Historical Journal 45, no. 2 (2002): 381–409; Hugh M. Coombs and John Richard Edwards, 'The Financial Reporting Practices of British Municipal Corporations 1835–1933: A Study in Accounting Innovation', *Accounting and Business Research* 25, no. 98 (1995): 93–105; E. P. Hennock, 'IV. Finance and Politics in Urban Local Government in England, 1835–1900', *The Historical Journal* 6, no. 2 (1963): 212–25.

[3] Anon., 'Opening of a Portion of the New Baths', *The Harrogate Herald*, 19 August 1896/NYCRO DC/HRG II 1–5 1891–1896 (reproduced in the minute books of Harrogate Corporation).

[4] 'Harrogate's Progress: A Big Rival to Continental "Cures", The Mecca of Society', *The Yorkshire Evening Post*, 15 September 1912/NYCRO DC/HRG/BPC 1911–1912.

[5] Anon., *Harrogate as a Health Resort: Published by the Corporation* (Harrogate: R. Ackrill, 1903), 22; Anon., 'Manchester at the Seaside: Prosperous Blackpool', *The Manchester Guardian*, 11 July 1896, 4/PQHN; Hayton R. Davis, *Harrogate as a Health Resort* (Harrogate: R. Ackrill, 1889), 10; Thomas H. Cole (ed.), *Hastings and St. Leonards-on-Sea as a Health and Pleasure Resort* (1884), 5. See also Richard Roberts, 'The Corporation as Impresario: The Municipal Provision of Entertainment in Victorian and Edwardian Bournemouth', in John K. Walton and James Walvin (eds), *Leisure in Britain 1780–1939* (Manchester: Manchester University Press, 1983).

[6] Coombs and Edwards, 'The Financial Reporting Practices of British Municipal Corporations 1835–1933', 97–8; Walton, *The English Seaside Resort*, 170.

achieved by running ads in numerous periodicals and magazines.[7] Advertising for holidaymaking was similar in essence, with just the relationship between producer and consumer switched around, given that the product was supplied chiefly to metropolitan consumers. By the late 1870s, the direct involvement of the corporations in advertisement had been firmly established.[8] Harrogate, which in 1914 was said to be 'the most conspicuous example in the country of successful advertising', had started strategically planning advertisements as early as 1868.[9] Blackpool corporation established a similar advertising committee in the late 1870s.[10] Systematic accounting of advertising expenses appear in the papers of Scarborough corporation only in the first years of the twentieth century, although there are signs that it undertook publicity campaigns earlier on.[11] In all three cases, the corporations' relation to advertisement was quite passive until the end of the century, and primarily consisted in considering offers made by independent agents and publishers.

The British advertisement industry went through a wave of professionalization in the 1880s, expanding coverage, developing new tactics of design and iconography, and giving rise to a large secondary industry of advice literature for advertisers.[12] In the late 1890s, municipalities began learning this professional approach to advertisement for themselves, discussing advertisement options in almost every meeting, sometimes by a specialized sub-committee. In 1897, Harrogate's 'Wells, Baths and Hackney Carriage Committee' changed its name to 'Wells, Baths, Establishment and Advertising Committee' and appointed a sub-committee to focus exclusively on advertising.[13] Later on, this new specialization led committees to employ designated professionals such as a 'Corporation Advertising Canvasser', artists, and printers, and they cooperated more actively than ever with external advertisement agencies.[14]

[7] Diana Hindley and Geoffrey Hindley, *Advertising in Victorian England, 1837–1901* (London: Hodder Wayland, 1972), 17.

[8] John K. Walton, 'The Re-Making of a Popular Resort: Blackpool Tower and the Boom of the 1890s', *The Local Historian* 24, no. 4 (1994): 194–205, here 199; Meeting of the Wells, Baths and Hackney Carriage Committee, 27 July 1875/NYCRO DC/HRG II 1/5 1875–1882.

[9] *The Daily Chronicle*, 14 May 1914/NYCRO DC/HRG/BPC 1914; NYCRO/DC/HRG II 1/5/1–1/5/7 1860–1891.

[10] Cross and Walton, *The Playful Crowd*, 27–8; Walton, 'The Re-Making of a Popular Resort: Blackpool Tower and the Boom of the 1890s', 199.

[11] Scarborough Corporation Minutes, Public Walks and Pleasure Grounds Committee 1890–1903, 17 October 1902/NYCRO DC/SCB II 1/13/1; Alfred Haviland, *Scarborough as a Health Resort: Its Physical Geography, Geology, Climate, & Vital Statistics, With Health Guide Map, &c.* (London: Hamilton, Adams and Co., 1883), v.

[12] Hindley and Hindley, *Advertising in Victorian England, 1837–1901*, 31–2. See for example: Anon., *The Advertiser's Guide to Publicity, etc.* (Birmingham: Moody's Printing Co., 1887); Francis David Lewis and Lowe, *The Advertiser's Guide to the Newspaper Press* [third edition] (London: Lewis and Lowe, 1844).

[13] Wells Baths etc. Committee, 15 November 1897/NYCRO DC/RIC II 1/5/9 1896–1900.

[14] Scarborough Corporation Minutes, Parks, Advertising and Entertainments committee 1909–1911, 2 March, 1910/NYCRO DC/SCB II 1/13/3; Wells, Baths, Establishment and Advertising Committee, 9 February 1909/NYCRO DC/HRG II 1/5/1–15 1906–1918.

In addition to the corporations and the advertising agencies, two other major players were directly involved in the resort industry: medical institutions and railway companies. Cooperation with health institutions (e.g. Board of Health, Medical Society) was naturally stronger in the spa resorts on account of them providing medical treatments. Spas needed medical authorization, consulting chemists in charge of ensuring the effective and safe supply of mineral waters, and therapists. But health institutions were also involved in the operation of seaside resorts, and their contributions were not necessarily limited to medical matters.[15] Interestingly, the medical profession was also deeply connected to the marketing work of the corporations. Some corporations published directly in medical magazines and journals.[16] In addition, numerous guides to holiday resorts were in fact written by physicians. Some reviewed multiple locations,[17] while others focused on one specific resort.[18] Despite claiming to be professional medical texts, these guides were no less part of the resorts' publicity and economy. The minutes of a meeting held by the 'Wells and Baths Committee' of Harrogate in May 1896 explicitly reveals this cooperation: 'It was left with a Sub. Committee', it is reported, 'to prepare & arrange for the insertion of 3 pages in Dr. Tiebel's Official Handbook of English & Foreign Spas at a cost of £25'.[19] The line between doctor and advertiser was blurred as corporation officials wrote the content of the entries on their resorts and doctors sold space in their well-known specialized books.

Medical knowledge and expertise took on a new, more complicated role as the resorts grew into commercial enterprises. Many popular guides and official booklets published by the corporations were compiled from anterior medical texts,[20] or included a recommendation of the resort signed by a certified doctor

[15] In Blackpool, for example, the Board of Health was responsible for the groundwork and construction of the sea front with its system of sea defences and a promenade (Walton, 'The Remaking of a Popular Resort: Blackpool Tower and the Boom of the 1890s', 199).

[16] Harrogate corporation published extensively in such platforms, as did to a lesser extent Scarborough corporation. Wells, Baths, Establishment and advertising Committee, 2 February 1899; 1 March 1900; 1 February 1901; 28 February 1901/NYCRO DC/RIC II 1/5/9 1896–1900; Scarborough Corporation Minutes, Parks, Advertising and Entertainments committee 1909–1911, 16 March 1910/NYCRO DC/SCB II 1/13/3.

[17] The most known being Dr Augustus Bozzi Granville's *The Spas of England, and Principal Sea-Bathing Places*, 1841; Dr. Burney Yeo's *Climate and Health Resorts*, 1890; and H. C. Tiebel, *Dr. Tiebel's Official Handbook of Spas, Watering Places, and Health Resorts* (London: Charles Letts & Co., 1896).

[18] For instance Davis, *Harrogate as a Health Resort* (1889); Cole, ed., *Hastings and St. Leonards-on-Sea as a Health and Pleasure Resort* (1884); Thomas F. I. Blaker, *Burgess Hill as a Health Resort: Hints to Those in Search of a Healthy Country Home* (Brighton: John Beal and Co., 1883).

[19] Wells Baths etc. Committee, 28 May 1896/NYCRO DC/HRG II 1-5 1891–1896.

[20] For instance Anon., *Harrogate. The Montepellier Estate. Proposed: Montepellier Spa, Baths, Concert Room, Winter Garden, and Hydropathic Company Limited* (Leeds: Inchbold & Beck, 1877)/HRG/PH/615.49; Anon., *Ward and Lock's Illustrated Guide to, and Popular History of Scarborough, Filey, Whitby, Btidlington Quay, and Their Vicinity* (London: Ward, Lock, and Co., 1880); Alfred Haviland, *Scarborough as a Health Resort: Its Physical Geography, Geology, Climate, & Vital Statistics, With Health Guide Map, &c.* (London: Hamilton, Adams and Co., 1883); Anon., *Visitors' Guide to Blackpool [Containing a Brief History of Blackpool, View of the Town, Street Plan and a Complete Guide*

as an 'additional and valuable testimony to the assertions made' in the publication.[21] Often, medical approval did not come as a formal endorsement, but through a general allusion to professional advocacy: '"Go to Blackpool and Live" say all the doctors who have lived since. And people have taken their advice and have come and gone, and come again and brought their friends with them, until to-day there is no more popular seaside resort in England than "The Brighton of the North".'[22] The quasi-mythological narrative in which the resort is said to have become popular because its medical properties were recognized by doctors appears in other guides, as for example in this Hastings publication:

> About a century ago, the position of Hastings, embosomed amid the surrounding hills, and guarded from every wind but the warm south-west, was noted by the medical men of that day as SINGULARLY SUITABLE FOR PULMONARY INVALIDS (original)—as, in fact, the Montpellier of England. They began to send their patients here, and the success attending the first cases induced the medical profession generally to recommend the place, which gradually increased in size and population.[23]

Like medical professionals and institutions, the railway companies also played a part in resort publicity. Much has been written on the role of the railway in the growth of tourism in Victorian Britain and elsewhere, and specifically on its impact on the development of the British holiday resort.[24] Indeed, the railway did much to assist the popularization of the holiday industry, not only in its function as a means of transportation, but also by promoting investment,

to All Places of Interest or Pleasurable Resort in Blackpool and the Fylde] (Blackpool: Blackpool Times, 1888), 61–2; W. MacDonald, *An Illustrated and Descriptive Account of Scarborough* (1896), 1–2/SCB; *Harrogate as a Health Resort* (1903) quoted an extract from an article on the reputation of British baths that was published in *The Lancet* a few years earlier (p. 18); Anon., *A New Pictorial and Descriptive Guide to Blackpool, Lytham, St. Anne's, Fleetwood, and The Fylde* (London: Ward, Lock & Co., 1902), 10.

[21] Anon., *Illustrated Guide to Blackpool and District* (Blackpool: S. Wilson & Son, 1891), 14. See, for instance *Harrogate: The Montepellier Estate* (1877)/HRG/PH/615.49; Cole, ed., *Hastings and St. Leonards-on-Sea as a Health and Pleasure Resort* (1884), 20–2; *Visitors' Guide to Blackpool*, 61–2; Davis, *Harrogate as a Health Resort* (1889); MacDonald, *An Illustrated and Descriptive Account of Scarborough* (1896), 1–2/SCB; Anon., *A New Pictorial and Descriptive Guide to Blackpool, Lytham, St. Anne's, Fleetwood, and The Fylde* (1902), 10; *The Cornish Riviera* (1906)/HAT 618/02/906/GRE, 13. Note that scientific approval was displayed both by seaside and spa resorts, cf. Durie, 'Promoting Spa and Seaside Resorts in Scotland c. 1750–1830', 197–8.

[22] *Visitors' Guide to Blackpool* (1888), 4.

[23] Cole, ed., *Hastings and St. Leonards-on-Sea as a Health and Pleasure Resort* (1884), 3.

[24] Pimlott, 'The English Seaside', 169; Walton, *The English Seaside Resort*, 26–9; Jack Simmons, 'Railways, Hotels, and Tourism in Great Britain 1839–1914', *Journal of Contemporary History* 19, no. 2 (1984): 201–22; Roy Millward, 'Railways and the Evolution of Welsh Holiday Resorts', in Anne K. B. Evans and John V. Gough (eds), *The Impact of the Railway on Society in Britain: Essays in Honour of Jack Simmons* (London: Routledge, 2016); Alastair J. Durie, 'Tourism and the Railways in Scotland: The Victorian and Edwardian Experience', in Evans and Gough, *The Impact of the Railway*.

planning, and construction in the tourism business.[25] British railway companies owned hotels and travel agencies, and their periodical literature (e.g. *Railway News*, *Railway Times*, *The Railway Magazine*), which provided information on technological advancements, traffic legislation, and negotiations between railway companies, also produced general knowledge on the tourism industry.[26]

The railway companies also expanded and improved the marketing of holiday resorts in cooperation with the local corporations: they served as publishing agencies and printed pamphlets and posters, they provided advertising space in their stations, and by the turn of the century, they had established regional advertising boards that allowed different resorts on a certain railroad track to collaborate with one another.[27] In some cases, the railway companies helped brand destinations. For example, as Jack Simmons notes, in addition to facilitating the journey between London and Southwest England, the railway branded Cornwall as 'The Cornish Riviera'.[28] In fact, during the first decade of the twentieth century, some railway companies founded their own advertising divisions.[29]

Railway companies and corporations also worked together to coordinate the timetables and rates for excursions to popular holiday destinations. The railways could rely on the corporations for information regarding the most attractive seasons, and the corporations could take advantage of the cooperation to control the 'social tone' of the visitors. The railway enabled increasing numbers of visitors to reach the resorts, but the resorts were deeply worried that the improvement of

[25] John Heeley, 'Planning for Tourism in Britain: An Historical Perspective', *Town Planning Review* 52, no. 1 (1981): 61–79, here 62; Simmons, 'Railways, Hotels, and Tourism in Great Britain 1839–1914'.

[26] Before the First World War, railway companies owned more than a hundred hotels throughout the United Kingdom. Additionally, railway companies bought travel agencies or designated their own travel agents (Thomas Cook being probably the most famous of them, appointed by the Midland Railway). Simmons, 'Railways, Hotels, and Tourism in Great Britain 1839–1914', 206, 210. On railway literature, see Edward Cleveland-Stevens, *English Railways, Their Development and their Relation to the State* (London: George Routledge and Sons, 1915), vii; G. A. Sekon, 'The Cornwall Railway Then and Now', *Railway Magazine* (1904); S. Kirkwood, 'Railway Tourist Literature', *Railway Magazine* 6 (1900): 510–6; Ian Yeoman, Alastair Durie, Una McMahon-Beattie and Adrian Palmer, 'Capturing the Essence of a Brand from Its History: The Case of Scottish Tourism Marketing', *Journal of Brand Management* 13, no. 2 (2005): 134–47, here 138.

[27] Roger Burdett Wilson, *Go Great Western: A History of GWR Publicity* (Newton Abbot: David & Charles, 1970); Anon., *Harrogate: Health, Happiness & Satisfaction* (GNR, 1913)/BL/GRC/YK.1992.a.5634; Anon., *Skegness Is So Bracing* (GNR, 1913)/BL/GRC/YK.1992.a.5634; Wells, Baths, Establishment and Advertising Committee, 28 January 1897/NYCRO DC/RIC II 1/5/9 1896–1900; Scarborough Corporation Minutes, Parks, Advertising and Entertainments Committee 1909–1911, 16 March 1910/NYCRO DC/SCB II 1/13/3; Wells, Baths, Establishment and Entertainments Committee, 14 February 1901/NYCRO DC/RIC II 1/5/10 1900–1906; Simmons, 'Railways, Hotels, and Tourism in Great Britain 1839–1914', 217; Heeley, 'Planning for Tourism in Britain: An Historical Perspective', 62–3.

[28] Simmons, 'Railways, Hotels, and Tourism in Great Britain 1839–1914', 214–5.

[29] Ibid. 217.

the means of transportation would draw the 'wrong' clientele.[30] Regulation of ticket prices was a practical way to control the social class of the visitors. As Simmons shows, the distinction between 'excursion', 'trip', and 'tour' reflected essentially social hierarchies.[31]

The cooperation between the corporations and the railway companies intensified after the turn of the century, when the corporations became more involved in advertising. In Scarborough, the chairman held meetings with 'the Chief Officials of the Publicity Departments of the GNR and the GCR, Messrs. Cook & Sons, Tourist Agents, and the Advertising Agents of the Underground Railways in London', and approved long-term marketing plans, advertising contracts, and annual advertising budgets.[32] Around the same time, the chairman of the 'Wells, Baths, Establishment and Advertising Committee' of Harrogate met the advertisement personnel of the GNR to coordinate the advertising strategy of 'Harrogate as a Health and Pleasure Resort'.[33]

Similar to the thin line between physicians and advertisers, the overlapping responsibilities of the corporation officials and the railway companies suggests that the resort was in many ways a shared business of the local government, the medical institutions, and the railway companies.[34] As we shall see, the negotiation between economic aspirations, classism, and medical theories among the different players in the holiday industry shaped its product. More than anything else, the unfamiliar frequent encounters between different social classes stood at the core of the shifting practices of holidaymaking and the related meanings of health and pleasure.

Responding to changes in clientele

The popularization of the British holiday market can be traced back to the mid-eighteenth century, when spa and seaside towns, already popular destinations for the aristocracy and gentry, started to attract members of the emerging middle

[30] Sue Farrant, 'London by the Sea: Resort Development on the South Coast of England 1880–1939', *Journal of Contemporary History* 22, no. 1 (1987): 137–62, here 140; Perkin, 'The "Social Tone" of Victorian Seaside Resorts in the North-West'.

[31] Simmons, 'Railways, Hotels, and Tourism in Great Britain 1839–1914', 209.

[32] Scarborough Corporation Minutes, Parks, Advertising and Entertainments Committee 1909–1911, 6 December 1909; 1 February 1910; 16 March 1910/NYCRO DC/SCB II 1/13/3.

[33] Wells, Baths, Establishment and Advertising Committee, 9 February 1909/NYCRO DC/HRG II 1/5/1-15 1906–1918.

[34] Indeed, *The Times* reported that the official event of the opening of new baths in Harrogate was attended by '[r]epresentatives of the medical profession from all parts of the country' and the invitation issued jointly by the GNR and the corporation, Anon., 'The Harrogate Baths', *The Times*, 4 July 1912/ NYCRO DC/HRG/BPC 1911–1912.

class, wealthy merchants, and tradesmen.[35] However, while this was a specific case of the assimilation of the nouveaux riches into 'polite society', the growth of the middle classes in the mid-Victorian years led to the development of a new model of resorts. These establishments were withdrawing from the aristocratic tone and social structure of the spa town and moving closer to the sea-bathing tradition of the lower classes, all while reflecting more of the particular tastes and mores of the urban middle classes.[36] A further transition took place between 1870 and 1890, when accessibility to the holiday industry accelerated, coming to include lower-middle-class members and manual workers.[37] The new diversity of visitors together with the resorts' explicit commercial objectives led the corporations to implement strategies of social marketing, whether by claiming to ensure a specific 'social tone' or arguing that they catered to all 'tastes'. Although all the resorts wished to attract higher numbers of visitors without risking their essentially middle-class ambience, as a matter of fact, distinctive holiday systems emerged. The distinction was made either in space (different resorts catering to a certain class) or in time (as the working classes were dependent on the bank holidays, the wealthy had more flexibility when choosing the time of their visit).[38]

The corporations' concerns about the changing clientele and the application of social marketing can be deduced from the distribution of advertisements and the content of their official guides. While some resorts, such as Brighton, Hastings, and Scarborough, managed to adapt to the changes in clientele and promote accommodation and entertainment for diverse markets, many smaller resorts catered mainly to middle-class families. Scarborough corporation was very successful in 'differentiated marketing': it invested in popular and local press as well as in cooperation with the railway companies, but it never lost its 'respectable' reputation.[39] In this way, it managed to help the working man, the excursionist, and the wealthy, 'each in his own style find health in his sickness, rest in his

[35] Wood, *Spa Culture and the Social History of Medicine in the Nineteenth Century*, 1; Walton, *The English Seaside Resort*, 6.

[36] Berghoff, 'Modern Tourism and the Rise of the Consumer Society'; John Soane, 'The Origin, Growth and Transformation of Maritime Resorts since 1840', *Built Environment* 18, no. 1 (1992): 12–26, here 14; Walton, *The English Seaside Resort*, 10–11, 25.

[37] Demands for working-class seaside holidays originated in the northern regions and especially in Lancashire, where a majority of workers could afford a full week holiday already in the 1880s, about a decade before their London peers and a quarter of a century before the working classes of South Wales (Walton, *The English Seaside Resort*, 25–26, 29–31, 35).

[38] Urry, *The Tourist Gaze*, 16; Soane, 'The Origin, Growth and Transformation of Maritime Resorts since 1840', 18; Walton, *The English Seaside Resort*, 23–25, 33; Farrant, 'London by the Sea: Resort Development on the South Coast of England 1880–1939', 138–40; John K. Walton, 'Residential Amenity, Respectable Morality, and the Rise of the Entertainment Industry: The Case of Blackpool', *Literature and History* 1 (1975): 62–78, here 64–5.

[39] Wells Baths etc. Committee, 28 July 1896/NYCRO DC/HRG II 1–5 1891–1896; Wells, Baths, Establishment and Advertising Committee, 15 November 1897; 2 February 1899; 1 March 1900/ NYCRO DC/RIC II 1/5/9 1896–1900; Scarborough Corporation Minutes, Parks, Advertising and Entertainments Committee 1909–1911, 16 March 1910/NYCRO DC/SCB II 1/13/3.

weariness, and occupation for his leisure',[40] while still being described by *Vanity Fair* as a fashionable destination that was 'neither desiring nor expecting the patronage of the ordinary "tripper"'.[41] Blackpool's advertising campaign also never gave up on the 'better class' market, although the resort was considered by the turn of the century as the exemplary working-class resort.[42] In a 1904 guide, it was presented as the ideal destination for 'the toiling millions of Great Britain' and as providing 'a rare combination of health and pleasure that more than compensates for the hard and remitting drudgery in factory, mine and workshop, and that has to be endured for the rest of the year'.[43] Yet, this straightforward appeal to the heart of the working classes was immediately balanced out with the claim that for 'the leisured masses with time and money at their disposal, there is nothing so good as a long stay at Blackpool, especially in the earlier and later periods of the year, when the crowds have thinned, and the discomforts of bank-holiday bustle have not to be endured'.[44]

Harrogate, on the other hand, decisively aimed to remain an exclusive resort and primarily published advertisements in media dedicated to the British 'high society' and aristocracy as well as in newspapers published in the colonies and in major Western cities.[45] Indeed, as a 1912 press article noted:

> The authorities at Harrogate have aimed high; they have set themselves to cater, not for all classes indiscriminately, but for the cream of society—for the highest in the land . . . Harrogate a year or two ago stood at the parting of the ways. On the-one hand were the elite, the glittering crowds which constitute English society . . . on the other the more mediocre multitude . . . Harrogate in the season is not for the crowd, not for the man in the street, not even for the mere rich man; it is out and out on the side of the aristocrat.[46]

[40] Anon., *Adey's Annual Popular Guide to Scarborough* (London: Alexander & Shepheard, 1887), 4; see also Anon., *Scarborough The Queen of Watering Places* (1914), 1–2/SCB.

[41] Anon., 'Scarborough – "Queen of Watering Places"', *Vanity Fair*, 25 June 1914/NYCRO DC/HRG/BPC 1914.

[42] Walton, 'The Re-Making of a Popular Resort: Blackpool Tower and the Boom of the 1890s', 199–200; Walton, 'Residential Amenity, Respectable Morality, and the Rise of the Entertainment Industry', 74; Walton, *The English Seaside Resort*, 33; Soane, 'The Origin, Growth and Transformation of Maritime Resorts since 1840', 18.

[43] Anon., *Blackpool Pictorial Guide* (Blackpool: J. P. Bamber, The Studio, 1904).

[44] *Blackpool Pictorial Guide* (1904).

[45] Harrogate advertisements were published, among other places, in *The Queen* / Wells, Baths, Establishment and Advertising Committee, 2 March 1899/NYCRO DC/RIC II 1/5/9 1896–1900; *The Court Circular*,/ibid. 2 August 1900; *The Washington Post*, *The New York Herald*, *The Chicago Daily Tribune*, *Times of India*, and *Le Figaro*/NYCRO DC/HRG/BPC 1911–1912; NYCRO DC/HRG/BPC, 1914.

[46] Anon., 'Harrogate's Progress: A Big Rival to Continental "Cures", The Mecca of Society', *The Yorkshire Evening Post*, 15 September 1912/NYCRO DC/HRG/BPC 1911–1912.

The social differentiation of holidaymaking was far from taking place only between individual resorts. In fact, resorts became sites of intense encounter between classes. Hence, a central task of social marketing was to address the fears and prejudices occasioned by such encounters. This was especially true of resorts like Blackpool, which were already acquiring a reputation as 'low class' destinations. The 1891 *Illustrated Guide to Blackpool and District* thus quoted an extract from the Blackpool *Times* stating that 'there is an entire absence of anything approaching vulgarity... and there is nothing of which the most fastidious could complain'.[47] Similarly, a decade letter, a different guide guaranteed that a 'Blackpool Sunday would prove a revelation to any who thinks slightingly of the town because it is the favourite resort of the toilers in the Lancashire mills', and continued to assure readers that the visitors are 'refined in manners'. 'The merriment of the town', the guide asserted, 'never descends to vulgarity; the crowd, though always joyous, is never boisterous; there is rarely to be seen even a trace of rowdyism to cast a blot upon Blackpool's good name'.[48]

Supplying health and pleasure

'Once upon a time', wrote an anonymous contributor to *The World* in 1912, 'life at a health resort consisted in drinking periodical small glasses of the "waters", and in walking off the effects. There was nothing else to be done.'[49] This statement is inaccurate. As a matter of fact, holiday resorts had been expanding their entertainment facilities throughout the nineteenth century. Just as inaccurate is the historiographic narrative describing amusement as supplanting the traditional objective of health and cure. In fact, entertainment facilities developed alongside the enduring cooperation between the resorts and medical institutions, and the advertisements and guide books of both spa and seaside resorts unequivocally touted the resorts' health benefits on the odd pages while portraying their amusement facilities on the even ones. When one takes the emotional component into consideration, it is evident that amusement did not simply substitute the resorts' traditional health-oriented objectives and practices; rather, amusement was in the process of acquiring a therapeutic meaning in and of itself. Indeed, this is precisely what the writer to *The World* explained in the second part of his piece: '[n]owadays things are very different', he wrote, because the authorities 'have wisely recognised that *amusement is necessary to health*; consequently visitors... find every facility for combining pleasure with the "cure"'.[50]

[47] Anon., *Illustrated Guide to Blackpool and District* (1891), 7.

[48] Anon., *A New Pictorial and Descriptive Guide to Blackpool, Lytham, St. Anne's, Fleetwood, and The Fylde* (1902), 16.

[49] Anon., 'Harrogate', *The World*, 12 July 1912/NYCRO DC/HRG/BPC 1911–1912.

[50] Ibid. My emphasis.

During the period discussed, and notably between 1890 and 1910, most British watering places operated under the slogan 'health and pleasure resort', which is to say that in addition to curative activities and treatments, they provided entertainment for their visitors, regardless if they were in ill health or not.[51] This combination was not at all new, as revealed by the eighteenth century descriptions of Scarborough as a town of 'health and merit' and of Bath as 'the resort of the sound, rather than the sick'.[52] Indeed, entertainment was an integral feature of spa culture and the seaside resort from the outset.[53] Yet, in the aristocratic water cure culture of the eighteenth century and the first half of the nineteenth century, the essence of the watering places was the cure, which was achieved by 'taking the waters'—drinking sulphur or sea water—and by participating in treatments such as bathing in baths or in the ocean.[54] The waters were usually taken before breakfast and the treatments were carried out during the daytime, meaning that 'taking the cure' only lasted a couple of hours and left most of the day free. Thus, resorts had to offer some kind of activities for their visitors in the long hours after their treatments as well as for water-takers' companions. The entertainment amenities usually consisted of an assembly room, a library, a newspaper room, a coffee house, and, later on, a theatre. The same facilities were adopted by the early seaside resorts, and were regarded, up through the early twentieth century, as the most basic and essential social institutions of any holiday resort.[55] However, until the late-nineteenth century, these amusements or attractions, as they were called, were not recognized as attractions in themselves, but as necessary supplements to the practices of 'taking the waters'. This is illustrated in the structure of late-eighteenth- and early-nineteenth-century guides, in which the 'amusements' section typically appeared towards the end, only after the detailed descriptions

[51] See, for instance Anon., 'The Traveler's Bureau: Harrogate, a Delightful Centre for Holiday and Cure Purposes', *The Gentlewoman*, 12 April 1912, 583/NYCRO DC/HRG/BPC 1911–1912; Anon., *Blackpool the Unrivalled Sea Side Resort for Health & Pleasure* (Blackpool: Blackpool Corporation, 1898), 49; Anon., *Adey's Annual Popular Guide to Scarborough* (1887), 21.

[52] Brodie, 'The first seaside resorts, 1730–1769', 22; Borsay, *A History of Leisure*, 25.

[53] Wood, *Spa Culture and the Social History of Medicine in the Nineteenth Century*, 1–9, 26; however, as Allan Brodie observes, '[e]ntertainment facilities in early resorts were initially small-scale and often rudimentary as visitor numbers were small and uncertain', Brodie, 'The First Seaside Resorts, 1730–1769', 24. See also ibid. 18; Walton, *The English Seaside Resort*, 157–9; Pimlott, *The Englishman's Holiday*, 36.

[54] Adams, *Healing with Water*; Noel G. Coley, 'Physicians and the Chemical Analysis of Mineral Waters in Eighteenth-Century England', *Medical History* 26, no. 2 (1982): 123–44; Borsay, 'Town or Country? British Spas and the Urban–Rural Interface', 156.

[55] Brodie, 'The First Seaside Resorts, 1730–1769', 22–4; Borsay, 'Visualising the Seaside, c. 1750–1914', 180; Borsay, 'Town or Country? British Spas and the Urban–Rural Interface', 156; Borsay, 'From Port to Resort', 91–2; Walton, *The English Seaside Resort*, 170, 173–5; Bryan Berryman, *Scarborough as It Was* (Nelson: Hendon Publishing Company Limited, 1971), 11; Pimlott, 'The English Seaside', 167; Pimlott, *The Englishman's Holiday*, 120; See also Durie, 'Promoting Spa and Seaside Resorts in Scotland c. 1750–1830', 197. It is worth mentioning that in contrast to their continental counterparts, entertainment facilities in British spas did not include casino gambling, see Walton, 'Spas in History, Spas and History', 10.

of treatments, the mineral composition of the waters, historical and geographical facts, church services, etc.[56]

In the mid-nineteenth century, new forms of organized entertainment began popping up at the resorts, including golf courses, tennis and cricket courts, and pleasure gardens; at the same time, seaside amusements were transferred to the beach, the promenade, and the pier.[57] This development concurred with the new scientific attention to the positive influence of nature described in the first chapter. We have seen that the medical literature increasingly emphasized the therapeutic potential of the emotional experience of nature, moving away from the previous focus on the benefits of the mineral composition of the waters and the related treatments and towards a general appraisal of the various pleasures to be experienced in the open air. This theoretical rationalization quickly infiltrated the public debate over the legitimate objectives of holidaymaking. Justifying the demand for 'mere pleasure', an author argued in a Manchester newspaper in 1899 that

> [t]here is a natural idea in the heads of most holiday-makers... that a time of holiday is really indulged in as being necessary to health rather than for pleasure. They go to places that they suppose are bracing, and talk meritoriously of inhaling the ozone of the atmosphere as if it were a chemists' draught... But the Englishman's affection for the sea as a good friend to the health of himself and his family is, after all, justified... The pleasure of being by, or at, or on, the sea is comparable to nothing else, and the russet ruddiness of complexion acquired is a true index of the health of body and of mind.[58]

The curative effect of nature was described in further detail in a 1901 piece published in *The Western Times*, in which it was claimed that even a single day 'in the open air' cleans the lungs, strengthens the blood, and bolsters the nerves. Like in some of the medical texts, both the remedy and the diseases described are generic, and the emphasis is on the generally beneficial effects of nature on health: 'The people get close to Mother Earth, and to her sea, and she instils her

[56] See, for instance Anon., *Thorpe's Visitors' Hand-Book for Harrogate, and Strangers' Companion to All Objects of Interest in the District: Together with a Popular Description of the Mineral Waters, and a Reduced Ordnance Map of 20 Miles round Harrogate, and Small Plan of the Town* (Harrogate: R. Ackrill, 1850); John Clifford, *The Tunbridge Wells Guide* (Tunbridge Wells: J. Clifford, 1837); Anon., *The Scarborough Guide* (Scarborough: T. Coultas, 1815); James Schofield, *The Scarborough Guide* (Scarborough: Thomas Lee and Co., 1796/SCB.

[57] Walton, *The English Seaside Resort*, 156–9; Farrant, 'London by the Sea: Resort Development on the South Coast of England 1880–1939', 138–9; Pimlott, 'The English Seaside', 166.

[58] Anon., 'Holidays and Health', *Supplement to the Manchester Courier*, 26 August 1899, 8/BNA. The mentioning of chemists as the former professional authority in holidaymaking at watering places hints directly at the medical context of the debate, which is analysed by Christopher Hamlin, 'Chemistry, Medicine, and the Legitimization of English Spas, 1740–1840', *Medical History Supplement* 10 (1990): 67–81.

subtle essences into jagged brains and wearied bodies, bracing both for another spell of labour.'[59]

A further refinement of the medical benefits of the holiday resorts was the increasing focus on the healing powers of air rather than of water. The idea of healthy air had been linked to the seaside holiday resort since its inception, but most of the mid-eighteenth and early-nineteenth century literature on the salubrious qualities of the resorts focused on the therapeutic benefits of the waters. This was to some degree an extension of the water cure culture of the spa, but also a new phenomenon related mostly to Dr Richard Russell's studies of the curative powers of drinking sea water. Gradually, and significantly with the growing institutionalization of climatological research, more energy was dedicated to promoting the seaside's general atmosphere, which was also a reaction to the emerging concern about the pollution of the growing urban centres. Thus, in the last quarter of the nineteenth century, a cult of sea air was having its day, and the curative properties attributed to the resort were transferred from the waters to the air, the sun, and the general climate.[60] 'Fresh air and sunshine', affirmed MD Charles Musgrove, a doctor specialized in nervous disorders, 'are the best restoratives for body and mind', and hence '[t]here is no form of bath equal to a sunbath'.[61] As analysed by John Beckerson and John K. Walton, resort publicity echoed the medical rhetoric and stressed the healthfulness of seaside air. Holiday resorts, they argue, were essentially 'selling air'.[62]

The healthfulness of seaside air, its embracing ozone and tonic, converted promenading into a central activity at every holiday resort, or, as Benjamin Brierley wrote in 1881 about a visit to Blackpool: '[w]e just did what lots besides us dun when they go'n to Blackpool *for their health* (original)... We paraded th' piers till everybody knew us.'[63] However, for some visitors, this kind of pleasure was unsatisfying. The gradual shift away from the traditional therapeutic functions of watering places and the increasing demand for amusement was met by some holidaymakers with expressions of sarcasm, boredom, and parodies of the medical theorization of holidaymaking.[64]

An anonymous article published in 1871 under the title 'Bored at the seaside' illustrates this attitude brilliantly. The text depicts a transitional stage in which although the declared objective of the holiday was still 'the cure', many visitors no

[59] Anon., 'Bank Holidays and Health', *The Western Times*, 28 May 1901, 4/BNA.

[60] Gray, *Designing the Seaside: Architecture, Society and Nature*, 21, 29; Beckerson and Walton, 'Marketing the Intangible at British Resorts', 55–6; Löfgren, *On Holiday: A History of Vacationing*, 116–18; Pimlott, 'The English Seaside', 166–7; Pimlott, *The Englishman's Holiday*, 106.

[61] Musgrove, *Holidays and How to Use Them* (1914), 30.

[62] Beckerson and Walton, 'Marketing the Intangible at British Resorts', 57.

[63] 'We just did what lot's besides us do when they go on to Blackpool for their health [...]' (all translations from dialect are mine). Ab-o'-th'-Yate [Benjamin Brierley], *Drop't on at Blackpool* (Manchester: Abel Heywood & Son, 1881), 17.

[64] Beckerson and Walton, 'Marketing the Intangible at British Resorts', 62.

longer took it seriously. The end result was that the resort did not yet offer them alternative activities that might better fulfil their desires:

> Having nothing whatever the matter with me, I am down at the sea-side for my health. Being naturally of a social, pleasure-loving turn, I ask eagerly that I may be amused. A period of enforced laziness is before me. The solicitations of my wife, and her cruel hints concerning the delicacy of our children, as well as her significant allusions to my own 'jaded look', and to the imperative necessity of our all 'having a change', have prevailed, and I am down here and down in the mouth as well. Three days have shown me pretty conclusively what my life will be for the next month, unless, indeed, this appeal is answered, and some form of amusement crops up . . . But my wife would have the local doctor in after the first morning if I ventured upon aught half so nice as this, and pills and boluses, grave warnings as to diet, regimen, and habits, followed by a long bill, would be the inevitable result. So I rise early, not because I like it, not because I consider it to be in the least degree necessary for my health, but for the reason that I regard it as the least evil of the two, and that being bored and wretched is better than being pilled and dosed.[65]

Such complaints became more frequent in the following decades.[66] In 1877, a reviewer claimed that 'at the ordinary English watering place there is little, if anything, beyond the pier and the beach; and from morning till night there is nothing to do but to sit on the sands or to walk up and down the pier . . . there is no place of entertainment or amusement'.[67] The boredom and lack of entertainment were often complemented by distrust and cynicism towards the curative powers of the holiday resort, as in this text written by someone who was sent to Torquay in order to cure a somatic condition:

> Yes; here I am, my editorial friend. Not of my own free will and accord, but by the advice of a medical practitioner whom my friends consider 'eminent'. All I can say is that his physic is as nasty as any I ever purchased at the chemist's shop—that the oil of the liver of the cod by which he orders me to be nourished makes me almost twist my head off my shoulders with nausea—and that his twaddle about 'regimen and diet' is eminently absurd. But I am admonished to be passive and obedient . . . 'Then', says my tormentor, 'you may have a change of air and scene'—a remedy, it strikes me, these eminent gentlemen always fly to when they are convinced they cannot cure you, and that if the patient *must* (original) die it is

[65] Anon., 'Bored at the Sea-Side', *The Observer*, 20 August 1871, 5/PQHN.
[66] See also Borsay and Walton, 'Introduction: The Resort–Port Relationship', 9.
[67] Anon., 'London, Sunday, July 1', *The Observer*, 1 July 1877, 4/PQHN.

> just as well—and perhaps better for them—if he performs that operation at a distance from home.[68]

This attitude was not only expressed by members of the middle classes. In two popular stories by Benjamin Brierley, who represents better than anyone else the Lancashire working-class experience of Blackpool, Brierley's alter ego character Ab-o'th'-Yate expresses similar sarcasm in sentences such as '[t]hey may co' Blackpool a healthy place, but there must be a good deeal o' widows theere',[69] or 'we fund they a lot o' folk theere ut had come-n moore for a frolic than for any good they expected to get eaut o' sae-wynt'.[70] Indeed, many visitors of watering places did not need or want 'the cure'. They went there looking for a social experience and entertainment. However, these accounts' contempt towards the medical discourse on the holiday and their juxtaposition of it with amusement is telling.

Edward Baker was a London chemist who visited Scarborough between September and October 1850. His diary makes it clear that he was more interested in the entertainment offered than in the water cure.[71] However, in contrast to later writings like those of the authors quoted above, who were familiar with the medical narrative but criticized and mocked it, the whole of Baker's diary contains not a single mention of health or treatments. Rather, the diary consists of a detailed account of the daily activities, which basically consisted in promenading, attending concerts and plays, and riding to nearby villages. This is what a typical day of Baker's visit looked like:

> Saturday, the 21st. A most tremendous boisterous day, we went out soon after breakfast to the Spa Saloon, to hear the German Band play, from 11 till 1 o'clock, and returned to the Hotel to the public luncheon. We afterwards took a stroll through some of the principal streets of the town to examine the shops, etc., and reached the Spa Terrace in time to witness the beautiful rough sea, at high tide, which came quite over the sea wall; remained watching the monster waves, until it was time to go in to dress for dinner, sat down about 70. After which we went a large party to the Theatre Royal, to see Sir E. B. Lytton's celebrated historical play called *Richelieu*; or *The Conspiracy* (original).[72]

[68] Anon., 'On My Back', *The Hull Packet and East Riding Times*, 25 June 1880/BNA.

[69] Ab-o'-th'-Yate [Benjamin Brierley], *Adventures at Blackpool* (Manchester: John Heywood, 1872), 14.

[70] 'They may call Blackpool a healthy place, but there must be a good deal of widows there; we found there a lot of folk that had come more for a frolic than for any good they expected to get out of sea-wind'. Ibid. 15.

[71] Edward Baker, *Sojourn in Scarborough 1850: The Diary of Edward Baker. With an Introduction by Patricia Fenteman* (Leeds: The Old Hall Press, 1984 [1850]).

[72] Ibid. 14.

Although most of his days seem quite similar and the resort's entertainment rather limited, in his entire diary Baker never mentioned experiencing boredom; as a matter of fact, he argued that 'much there is to amuse and interest a visitor in this sea port town, which will occupy quite a long day'.[73] In contrast, the anonymous author writing in 1871 quoted above, being 'sick of cribbage, backgammon, draughts, and double-dummy', concluded his report declaring: 'if I am ever caught booking myself at an English watering place for a whole month again, may I be punished as I deserve'.[74]

As sociologist Jack Barbalet claims, the experience of boredom is significant because it points to a rupture in the consensual 'interpretation of the meaningfulness or otherwise of the activity or situation in question'.[75] Expressions of boredom are thus extremely valuable for the historian, as they might provide a clue about a moment of change. In the world of British holidaymaking, the feeling of boredom seems to have anticipated an essential transformation in the holiday experience. The impact of vacationers' boredom (or at least their choice to express it) can also be deduced from resorts' repeated promises that their visitors could enjoy plenty of amusements and would experience no dullness whatsoever.[76]

During the 1870s and 1880s, the practice of promenading in the embrace of the sea air and the sounds of the orchestra was complemented by commercial forms of entertainment, including black minstrel and pierrot shows, food vendors, shops, and, later on, slot machines.[77] Gradually, the 'pleasure seeker' became an accepted phenomenon, and the provision of entertainment and attractions for the visitors was seen as the corporation's responsibility.[78] While claiming that 'a sea-side resort should above all things be amusing'[79] was not overly common in the 1870s, from the 1890s on, with the increasing involvement of commercial companies in the entertainment business, amusement was on the ascendant as an independent objective for holidaymaking. Advertisements underscored the recreational facilities of resorts, and in some cases the description of the resort's

[73] Ibid. 7. [74] Anon., 'Bored at the Sea-Side', *The Observer*, 20 August 1871, 5/PQHN.

[75] Jack M. Barbalet, 'Boredom and Social Meaning', *The British Journal of Sociology* 50, no. 4 (1999): 631–46, here 636.

[76] Haviland, *Scarborough as a Health Resort* (1883), 7–8; *The Daily Graphic*, 20 July 1912/NYCRO DC/HRG/BPC 1911–1912. There are numerous similar examples; see for instance Haviland, *Scarborough as a Health Resort* (1883), 7–8; Anon., *Seaside Watering Places: A Description of Holiday Resorts on the Coasts of England and Wales, the Channel Islands, and the Isle of Man, Including the Gayest and the Quietest Places* (London: L. Upcott Gill, 1897), 323; Anon., *Blackpool the Unrivalled Sea Side Resort* (1898), 39; Anon., 'At Harrogate', *The Gentlewoman*, 19 November 1912/NYCRO DC/HRG/BPC 1911–1912.

[77] Walton, *The English Seaside Resort*, 165; 167–70; 173–5.

[78] Roberts, 'The Corporation as Impresario'; Charles H. Mate and Charles Riddle, *Bournemouth: 1810–1910. The History of a Modern Health and Pleasure Resort* (Bournemouth: Messrs. W. Mate & Sons, Ltd., 1910), 134–5; Anon., 'London, Sunday, July 1', *The Observer*, 1 July 1877, 4/PQHN; Anon., 'Bored at the Sea-Side', *The Observer*, 20 August 1871, 5/PQHN.

[79] Anon., 'Bored at the Sea-Side', *The Observer*, 20 August 1871, 5/PQHN.

leisure facilities outshined the conventional sections on climate, minerals, and curative powers.[80]

The corporations rejoiced about the superior profit potential of the growing popularity of commercial amusements over that offered by the mere enjoyment of nature. When a *Manchester Guardian* journalist went to investigate the source of '[t]he prosperity of Blackpool', described in the article as 'a wonder of the age', this is what he found out:

> 'We ourselves are a bit astounded by it', was the admission of one leading citizen to whom a representative of this journal was introduced the other day, towards the close of an investigation into the facts of the case. 'I can scarcely tell you', he added, 'why we are so popular.' 'Your weather, perhaps,' I suggested; 'last year, you know, you had about fifteen hundred hours of bright sunshine.' 'Had we, indeed?' he replied. 'No, I don't think it's the weather altogether that does it.' Nor was this gentleman content to give the sea the credit, although he was willing enough to admit that that element was of some importance to a seaside resort. It was from the lips of the most remarkable official in Blackpool that a quite unhesitating explanation eventually came. 'Our weather', said he, 'is as good as the best, and our sea *is* (original) the best, but if you want to know what has really made Blackpool you must look at our amusements. Our amusements are not to be beaten anywhere. You could be here a month and find a new amusement every day.'[81]

All this is not to say that 'pleasure' had substituted 'health'. Up until the turn of the century, most guides still proclaimed health to be the core objective of a visit to the resort. Guides to spa and health resorts with a long 'cure' tradition still included information for the ill and specified the diseases particularly benefited by a visit to their establishments, which were generally comprised of somatic, nervous, and emotional conditions. Inland spas as well as seaside resorts continued to stress their curative powers, detail the minerals of their waters, and provide instructions for 'taking the cure'.[82] Up until the 1890s, even a popular seaside resort like Blackpool could be described as a 'typical health resort', and its corporation's

[80] 'Health, Recreation & Amusement', *The Pioneer*, 24 July 1914/NYCRO DC/HRG/BPC 1914; 'Health for the Sick: Amusement and Sport for the Strong', *The Pioneer*, 24 July 1914/NYCRO DC/HRG/BPC 1914; 'For Good Golf and Amusement Without Fatigue Visit The Nation's Spa, Harrogate', GNR/NYCRO DC/HRG/BPC 1914. See also Farrant, 'London by the Sea: Resort Development on the South Coast of England 1880–1939', 142.

[81] Anon., 'Manchester at the Seaside: Prosperous Blackpool', *The Manchester Guardian*, 11 July 1896, 4/PQHN.

[82] See for instance Cole, ed., *Hastings and St. Leonards-on-Sea as a Health and Pleasure Resort* (1884), 3; *Harrogate as a Health Resort* (1903), 8; Anon., *Ward and Lock's Illustrated Guide to, and Popular History of Scarborough, Filey, Whitby, Btidlington Quay, and Their Vicinity* (1880), 12; Anon., *Adey's Annual Popular Guide to Scarborough* (1887), 20. See also Borsay, 'Visualising the Seaside, c. 1750–1914', 180.

guide of 1898 declared that '[t]he healthiness of Blackpool needs no wordy colouring', assuring, for example, that '[m]uscular, nervous and pulmonary diseases are cured as if by magic'.[83]

The first decade of the twentieth century appears to mark the transition point in the status of holiday resort amusements, which could now be regarded as a central, independent, and legitimate objective of holidaymaking. This transition is nicely depicted in contemporary popular literature. Let us consider for example two stories about Blackpool: 'Drop't on at Blackpool', another story by Benjamin Brierley featuring Ab-o'th'-Yate written in 1881, and Arthur Laycock's *Fun on the Pleasure Beach* from 1909. Brierley's narration begins as follows:

> After a painful illness o' two days, caused by havin' a damp throat an' a limestone stomach, I've bin eaut again, an' catcht a bit moore o'th' disorder... Sam Smithies said he thowt Blackpool would do me good... at anyrate, i' this my latest do at Blackpool.[84]

This is a common 'water cure' narrative: Ab is ill and is therefore advised to visit the seaside. A couple of pages further we find a rhyme about Blackpool. After the opening sentence: 'If yo' wanten an' eaut, or a spree, or yo'r health',[85] the poet describes the highlights that make Blackpool a favourite seaside resort. Among these are mentioned the donkey rides and 'drinkin' an' dancin'' at Uncle Tom's Cabin, one of the most well-known institutions of Blackpool. However, most of the rhyme illustrates the pleasures of nature: the sun, the sea and its waves, and the crowded Promenade.[86] Almost thirty years later, a similar poem included in Laycock's *Fun on the Pleasure Beach* displays a sharply different ratio between nature and artificial amusements. Among many other forms of entertainment at 'Pleasure Beach' Laycock mentions the skating rink, the scenic railway, the water chute, and the velvet coaster. Health, on the other hand, is not mentioned at all, and certainly not as the primary motivation for the holiday.[87]

Towards the First World War, only traditional spas were still detailing the curative powers of their treatments, and it is not unusual to find guides to Scarborough or Harrogate stating that 'the majority of people to-day do not

[83] Anon., *Blackpool the Unrivalled Sea Side Resort* (1898), 25–7; see also Walton, 'The Re-Making of a Popular Resort: Blackpool Tower and the Boom of the 1890s', 200; Anthony Hern, *The Seaside Holiday: The History of the English Seaside Resort* (London: The Cresset Press, 1967), 10.

[84] 'After a painful illness of two days, caused by having a damp throat and a limestone stomach, I've been out again, and caught a bit more of the disorder... Sam Smithies said he thought Blackpool would do me good... at any rate, this is my latest do at Blackpool'. Ab-o'-th'-Yate [Benjamin Brierley], *Drop't on at Blackpool* (1881), 3.

[85] 'If you want an out, or a spree or your health'. Ibid. 6.

[86] 'drinking and dancing'; 'the westerly sun'; 'the waves playing leap-frog like school lads'; 'the waves like wild horses'; 'bathing in the water'; 'the Promenade crowded with folk'. Ibid. 6.

[87] Arthur Laycock, *Fun on the Pleasure Beach wi' Sally an' Sam, Jimmy an' Jane Ann, towd by Sally hersel. A Funny Pennorth* (Blackpool: Laycock & Co., 1909).

regard the Spa so much as a resort for "taking the waters"',[88] or admitting that '[m]any of us start out in search for health at one time or another, most of us in search of pleasure'.[89]

Amusement and class

The development of artificial amusement as an independent objective for holidaymaking was not welcomed by everyone, and like other emerging spaces of consumerist leisure, the holiday resort embodied ongoing debates about 'proper' forms of recreation, idleness, class, and morality.[90] Under these social circumstances, the types of amusement offered became a proxy for the respectability of a resort, given that the deviation from the traditional characteristics of the health resort opened it to charges of moral degeneracy. Thus, not unlike urban forms of 'rational recreation', holiday resorts were considered by some to be 'fit sites for educational purposes', in which the social mixing could promote the 'improvement' and moralization of the lower classes.[91] This provided the grounds for forbidding certain forms of entertainment, such as cock fighting and alcohol consumption.[92] However, the essential debate was not about particular practices, but rather on the legitimization of holidaymaking for the lower classes. For many, acknowledging that holidaymakers from the higher strata distanced themselves from the traditional curative practices of watering places was one thing, but accepting that the lower classes could also enjoy the privilege of a seemingly hedonistic experience was different. Thus, although most middle-class visitors to the seaside gradually engaged more in amusement and idleness than in strictly health-related practices, the health-versus-amusement controversy reflected a wider social debate.[93] The vacation industry, which was already addressing the issue of social class through strategic marketing, had to adjust itself to the shifting social meanings of amusement. As Walton rightly notes, some resorts wishing to appeal to 'respectable' middle-class visitors inflated the differences between the practices of holidaymaking. Thus, the contrast between shows, 'artificial gaiety',

[88] Anon., *Scarborough The Queen of Watering Places* (1914), 1–2/SCB.

[89] Anon., 'The Traveler's Bureau: Harrogate, A Delightful Centre for Holiday and Cure Purposes', *The Gentlewoman*, 12 April 1912, 583/NYCRO DC/HRG/BPC 1911–1912.

[90] Gary Cross, 'A Right to be Lazy? Busyness in Retrospective', *Social Research* 72, no. 2 (2005): 263–86; Lynn M. Voskuil, 'Feeling Public: Sensation Theater, Commodity Culture, and the Victorian Public Sphere', *Victorian Studies* 44, no. 2 (2002): 245–74, here 251–2; Huggins, 'Leisure, Respectability and the Male Middle Classes in Victorian England'.

[91] Haviland, *Scarborough as a Health Resort* (1883), 7–8.

[92] Walton, *The English Seaside Resort*, 4, 162–3.

[93] See also ibid. 166–7; Walton, 'The Re-Making of a Popular Resort: Blackpool Tower and the Boom of the 1890s', 198.

and excitement on the one hand, and health, relaxation, nature, and quiet, on the other, was emphasized and reiterated in turn-of-the-century guides.[94]

The moral overtones of the health/amusement question were obviously dependent on existing assumptions about both health and amusement. But what if amusement was to be regarded as healthful in and of itself? Indeed, just as it had served as a powerful tool for justifying holiday legislation, therapeutic rationalization also played a central role in the social debate on the issue of 'reasonable amusements'. Physicians of the 1880s and 1890s were aware of the trends in demand and of the distancing of the resort industry from its traditional medical functions, and acknowledged that 'holiday plans are made with a view both to health and pleasure'.[95]

Some considered recreation and entertainment to be necessary components of the medical treatment, whether because a bored visitor might 'crave to leave the resort before his health is established', or because they believed that an active mind benefited the physical cure.[96] But in the 1890s, physicians began ascribing amusement its own independent salubrious effects, going beyond framing it as merely contributing to medical treatment. The impact of the holiday on one's health was increasingly presented as being dependent on a complex system of causes, in which amusement and entertainment had a central and beneficial role.[97] This scientific theorization of entertainment and commercial amusement was closely related to the medical conceptualization of emotions. We have seen that from the late 1880s, the medical literature began incorporating the new category of emotion into the existing philosophy of 'change', thus emphasizing the value of emotional alternation, including that acquired from artificial amusements. Once certain emotional states were pathologized, it became reasonable to conceive of counter-emotional experiences as curative. Holiday-as-change was therefore not only about a change of place, but also a change of emotional atmosphere. The pathologization of emotions and the related negotiations of the meaning of health and healing enabled a shift in the role of amusement from merely a supportive means for the therapy to the therapy itself.

This process was politically charged precisely because it was closely related to the emotional pathologization of 'the worker'. Let us return to the 1911 debate

[94] Anon., *Seaside Watering Places* (1897), 54–7, 167, 185; Anon., *A New Pictorial and Descriptive Guide to Blackpool, Lytham, St. Anne's, Fleetwood, and The Fylde* (1902), x; *Ward Lock Guide to Torbay*, quoted in Walton, *The English Seaside Resort*, 166.

[95] Squire, 'The Sulphur Springs of Great Britain and Their Therapeutic Action' (1888), 201.

[96] Haviland, *Scarborough as a Health Resort* (1883), 57–8; see also C. B. Keetley, 'On the Time and Place of Bathing and on the Hotel and Boarding-House Dietary at Health Resorts', *The Journal of Balneology and Climatology* 1, no. 3 (1897): 222–4.

[97] Yeo, *Climate and Health Resorts* (1890), 252, referring to Davos in Switzerland; Anon., 'As Others See Us', *The Journal of Balneology and Climatology* 1, no. 3 (1897): 291–94, 293 citing an article published in the *London Daily Telegraph* on 27 February 1896; Anon., 'British Balneological and Climatological Society', *The Lancet* 1, no. 3799 (1896): 1723; Anon., 'The Claims of British Spas', *The Lancet* 2, no. 4752 (1914): 804.

that a 'London Physician' provoked in Harrogate: we can now better understand the doctor's suggestion that the amusements 'outside the actual baths buildings' be improved. The 'London Physician' represented many other professionals of the time who viewed the therapy offered at holiday resorts as not being confined to the water treatments, but who rather saw the experience of the holiday as a whole as curative. From this perspective, we can also better understand the strong rejection of his advice by the mayor and other members of Harrogate corporation.

While the mayor ridiculed the 'London Physician' by maligning 'the joy-wheel form of attraction' as a kind of practice 'for the public on the lines of Blackpool or Morecambe'[98]—popular working-class seaside resorts—the chairman of the 'Wells and Baths Committee' of Harrogate, Councillor Sheffield, adopted a different approach. Instead of rejecting the Londoner's critique, Sheffield argued that the 'London Physician' was completely right, but was saying nothing new. 'Harrogate was now', he said, 'where A London Physician thought it would be in ten years' time if his advice were followed.' Moreover, Councillor Sheffield strongly rejected the notion that Harrogate lacked sufficient attractions: 'What about the Kursaal', he asked, 'the theatre, the Valley Gardens, the Moor, the Stray?'[99]

Harrogate was one of the most exclusive spa resorts in England and was frequently visited by members of the monarchy. Since the pathologization of emotions was now primarily focused on workers, the health-versus-pleasure question was a site of intense class struggle, as it legitimized members of all classes to enjoy the same privileges of 'pleasure'. It is therefore understandable that the aristocratic spa rejected the amusements of the working-class seaside resorts as a symbol of inferior social habits. However, as amusements were an integral part of the traditional spa culture as well, the debate in fact revolved around the 'appropriate' amusements. Indeed, as early as 1896, in the ceremony of the opening of the 'Victoria Baths' in Harrogate, the mayor, Charles Fortune, explicitly referred to the question of amusements, with the *Harrogate Herald* reporting that 'he thought they should keep the ball rolling, and provide legitimate entertainment for the visitors'.[100] In 1911, a commentator from Harrogate responding to the 'London Physician' argued that 'Merry-Go-Rounds and scenic railways are not exactly the best things for an invalid's nerves', while noting that '[a]s for Music and reasonable amusements, Harrogate is well provided'.[101] Expressions such as 'legitimate' and 'reasonable' expose the fact that leisure in general, and amusing

[98] Anon., 'No Joy Wheels Wanted: Mayor of Harrogate's Reply to Medical Critic', *The Yorkshire Evening Post*, 30 September 1911, 5/NYCRO DC/HRG/BPC 1911–1912.

[99] Ibid.

[100] Anon., 'Opening of a Portion of the New Baths', *The Harrogate Herald*, 19 August 1896/NYCRO DC/HRG II 1–5 1891–1896 (reproduced in the minute books of Harrogate Corporation).

[101] V. C. Mallan, 'Is Harrogate Enterprising? Visitor Thinks Council Should "Let Well Alone": The Rest Cure and Merry-Go-Rounds', *The Yorkshire Evening Post*, 12 October 1911/NYCRO DC/HRG/BPC 1911–1912.

recreation in particular, were a 'contested cultural space' in which notions of class, gender, and social respectability were debated.[102]

The past decade has witnessed an unprecedented interest in the relation between health and pleasure in holiday cultures.[103] However, most of these studies fail to provide a comprehensive explanation for the connection between the two concepts and to address the wider context within which this connection materializes, namely the relation between leisure, class, gender, scientific expertise, and emotion knowledge. Essentially, 'health' and 'pleasure' are still perceived as detached opposites. Referring to the 'balance' between therapy and pleasure in medical travel, George Weisz writes: 'In some cases I can say definitively that one or the other dominates. In many other cases, they coexist in permanent symbiosis or tension.'[104] In the case of British spa and seaside holiday resorts, publicity material and medical texts around the turn of the century show that this tension was closely linked to the interrelation between medical knowledge and class values. Moreover, it suggests that 'health and pleasure' should not be understood as the sum of two independent and a-priori defined concepts. As a matter of fact, the meaning and value of both concepts shifted in reaction to the complex relations between contemporary ideals of therapy and morality and the commercial aspirations of the holiday industry. A close examination of the late-nineteenth century resort demonstrates not only how hard it is to differentiate between commercial and political or between advertisement and professional review, but also how fragile the line between 'healthy' and 'moral' really is. While health was seen as a scientific, biological, and (relatively) universal issue, modes of recreation were a matter of taste, and were therefore attached to class and manners. But once amusement was framed as a practice aimed at curing an allegedly scientific pathology of the emotions, the social definitions of the purpose of recreation could slowly be altered.

The comparison of resorts that catered to visitors from different social classes also demonstrates that the justification of the holiday-for-fun cannot be interpreted as a 'top-down' process, as most emulation theories applied to tourism have done.[105] The controversy around the letter sent by 'A London Physician' well

[102] Huggins, 'Leisure, Respectability and the Male Middle Classes in Victorian England'.

[103] To mention a few examples: In *Healthy living in the Alps*, Susan Barton explores the seemingly surprising fact that Alpine resorts were frequented both by invalids suffering from lung diseases and by healthy winter-sports lovers in Alpine resorts. Susan Barton, *Healthy Living in the Alps: Sanatoria and the Origins of Winter Sports Tourism in Switzerland, 1860–1914* (Manchester: Manchester University Press, 2009); Diane Koenker, focusing on the Soviet context, traces the shift from the *otdykh* form of vacation associated with healthy rest to *turizm*, a more playful and consumerist type of vacation. Koenker, *Club Red: Vacation Travel and the Soviet Dream*; Recently, Marina Soroka also refers to the 'therapy versus pleasure' theme, dividing nineteenth-century European spas between 'serious spas' and 'post-cure spas'. Marina Soroka, *The Summer Capitals of Europe, 1814–1919* (Oxon: Routledge, 2017).

[104] Weisz, 'Historical Reflections on Medical Travel', 138.

[105] Berghoff, 'Modern Tourism and the Rise of the Consumer Society'; Campbell, 'Conspicuous Confusion'.

illustrates the influence of the working-class seaside resorts on the aristocratic health resorts. Furthermore, the analysis of material from both spa and seaside resorts shows that the categorical distinction between the two institutions is misleading. Different types of resorts had much in common, precisely because they were operating under the same social conditions in an era that saw the commercialization of the resort industry, its increasing accessibility and social diversity, and paradigm shifts in the medical sciences. The commercialization of the resort, which is to say, its move towards becoming a profit-oriented institution, had a powerful impact both on the social question (giving it an incentive to attract more visitors) and on the role of medical institutions (by using scientific knowledge as marketing material). In this emotional economy, as will be further explored in the next chapters, the roles of the producer, the advertiser, and the doctor overlapped, as they were all at once responsible for producing the meaning of 'the emotional' as well as the experience of it during the holiday.

4
The emotionalization of the holiday resort

The preceding analysis of the ways in which scientists, politicians, moralists, and the operators of the resort industry formalized the objectives of the holiday has laid bare the different layers of the emotional economy of holidaymaking: the construction of emotions as objects that can be pathologized, desired, and regulated; the social and political legitimation of time designated for the experience of positive emotions; the establishment of holidaymaking as a commercial leisure industry. We can now focus on the supply of emotional experiences, their conceptualization, and their production. Whereas the framing of holidaymaking and holiday amusements in emotional terms in the medical literature can be traced back to the 1880s, it was only in the first decades of the twentieth century that this was reflected in the marketing literature of the holiday industry. Alongside justifying amusement, spa and seaside resorts proclaimed the emotional effects of holidaymaking and gradually held up the emotional experiences of happiness and joy as the products being supplied.

This process did not escape the eyes of contemporaries, as can be seen in this extract from the 1902 *Ward Lock* guide to Blackpool:

> When there were no palaces of pleasure in Blackpool, and no electric trams; when the only promenade was an unkempt, unpaved sea-bank; and the not too dizzy delights of the place found their climax in the mild pleasures of Uncle Tom's Cabin . . . the town was still a favourite resort of those who sought relief from nervous complaints and pulmonary diseases. Tired-out workers, whether mentally or physically, were sent to be braced up by the air along this coast, and the place quickly grew in repute as a health resort. Then the municipal authorities saw their opportunity, and by degrees built, lengthened, and widened the promenade; the electric trams came, and with them the pioneers of the superb palaces of pleasure; and at last the favourite health resort, with still the same recuperative quality of air, became a veritable Fairyland of Delight.[1]

This description precisely identifies the shifts undergone by the holiday resort's product: from a 'health resort' that did not offer many entertainment options, to the development of the promenade, where visitors could 'be braced up by the air',

[1] Anon., *A New Pictorial and Descriptive Guide to Blackpool, Lytham, St. Anne's, Fleetwood, and The Fylde* (1902), 9.

The emotional economy of holidaymaking: Health, pleasure, and class in Britain, 1870–1918. Yaara Benger Alaluf, Oxford University Press (2021). DOI: 10.1093/oso/9780198866152.003.0005

and finally to the growth of the 'palaces of pleasure', providing a 'fairyland of delight' to its visitors. This last stage—the emotionalization of the resort—is the focus of the present chapter.

Advertising emotions

Scattered references to emotions can be found in advertisements before the turn of the century, although they are usually intertwined with health. Thus, whereas in 1898 there was 'not a healthier nor a happier spot'[2] on earth to be found than Blackpool, in 1902 the resort is described as a 'merry holiday resort', where 'the happiness of the place seems to grow with the density of the throng', which is 'always joyous'.[3] In the early twentieth century, emotional descriptions gradually came to dominate the literature on and advertising for resorts. Referring to Margate, a 1912 guide stated that 'a full measure of relaxation and pleasure can be extracted from a holiday spent at this deservedly popular seaside resort'.[4] A guide to Skegness mentioned that the resort afforded its visitors 'seaside pleasures and jollity'.[5] It expounded on the direct influence of nature on one's emotions, saying that on Skegness's 'sunny sands everyone feels the joy of living. The sea breezes blow away every cobweb of care from the brain, and under the influence of contagious happiness the wrinkles caused by age and worry are smoothed away.'[6]

A 1914 article declared that 'Scarborough is during the summer season the gayest of the gay.'[7] A different guide of the same year guaranteed that 'Scarborough in the evening is the same happy place as it is in the height of the day',[8] and that it offers 'something to harmonise with your every mood, and everything necessary for ensuring an ideal holiday'.[9] Even Harrogate, which still continued to stress the healing 'medicinal properties of its waters' and treatments, was using emotional language in its ads: 'Harrogate: Health, Happiness & Satisfaction'; 'Harrogate – for a healthy, happy holiday'; 'for Health and Enjoyment – HARROGATE'.[10] The new significance of emotional experience as

[2] Anon., *Blackpool the Unrivalled Sea Side Resort* (1898).

[3] Anon., *A New Pictorial and Descriptive Guide to Blackpool, Lytham, St. Anne's, Fleetwood, and The Fylde* (1902), 16.

[4] Anon., GNR booklet (GNR, 1914)/BL/GRC/YK.1992.a.5634.

[5] Anon., *Skegness Is So Bracing* (1913), 1/BL/GRC/YK.1992.a.5634.

[6] Ibid. 2.

[7] Anon., 'A Holiday Highway – An Answer to an Annual Question: The Great Northern Railway Route to Scenes of Interest and Beauty', *The Gentlewoman*, 4 July 1914, 24 / NYCRO DC/HRG/BPC 1914.

[8] Anon., *Scarborough the Queen of Watering Places* (1914), 34 / SCB.

[9] Ibid. 1–2.

[10] *Harrogate: Health, Happiness & Satisfaction* (1913), 2; 'Harrogate – For a Healthy, Happy Holiday', *North Mail*, 22 June 1914 / NYCRO DC/HRG/BPC 1914; 'Charming Resort in England for Health and Enjoyment', *Town & Country*, 6 May 1912 / NYCRO DC/HRG/BPC 1911–1912.

a central quality in the description of a resort even allowed a journalist to call Harrogate, an inland spa resort, '[a] cheerful seaside town'.[11]

Together with the emotional language, a new element appeared in the resorts' publicity: holiday memories. 'Taking the waters' was known to be an unpleasant experience that consisted of despair, pain, and traumatic treatments. Seeking to forge a contrast, advertising for the emerging emotional holiday touted a different type of advantageous, long-lasting effect:[12] the 'pleasant' feelings enjoyed during the holiday could be reactivated by simply mentioning the name of the resort: 'Scarborough! What pleasant memories the name recalls to those who have already spent a holiday by its wave-beat shore.'[13] The affinity between emotion and memory stands at the core of Colin Campbell's influential work *The Romantic Ethic and the Spirit of Consumerism*. In his distinction between traditional and modern hedonism, Campbell argues that, first, the modern hedonist essentially seeks emotional pleasure and not merely sensorial pleasures, and second, that this emotional pleasure is fulfilled through imagination or daydreaming.[14] This practice of fantasizing, Campbell stresses, differs from traditional desire in that it can evoke 'future experiences . . . which one has had no experience of at all'.[15] This is exactly what the second part of the sentence extracted from Scarborough's 1914 official guide implies: 'Scarborough! What pleasant memories the name recalls to those who have already spent a holiday by its wave-beat shore *and what a treat there is in store for those who have not as yet come under its magic spell.*'[16] Similar rhetoric was used in this 1915 advertisement: 'Blackpool! The seaside! The sweet, bright, happy, exhilarating, life-renewing seaside! What a wealth of grateful and joyous memories spring into being at the sound of the name.'[17]

Campbell's thesis was further developed by Eva Illouz in her analysis of the relation between imagination and emotions in consumer culture. Intending to identify the mechanisms behind Campbell's theory, Illouz argues that through the

[11] Anon., 'Holidays for the Nervy', *Daily News & Leader*, 10 April 1914 / NYCRO DC/HRG/BPC 1914. This is an inaccurate quotation from page 46 of Charles Musgrove's book *Holidays and How to Use Them* (1914). The original sentence is: 'A cheerful seaside town or an inland resort, such as Buxton or Harrogate, is often better suited to them.'

[12] On the unpleasantness of the water cure, see Steward, 'The Affective Life of the Spa', 216–9; Pimlott, *The Englishman's Holiday*, 57. The negative emotions associated with the water cure contradict George Weisz's claim that 'the therapy and the pleasures of tourism are inextricably linked and are intimately associated with specific places considered especially "healthy" and, not inconsequentially, pleasant and attractive' (Weisz, 'Historical Reflections on Medical Travel', 138). What Weisz regards as 'inconsequential' must be validated empirically for each specific case.

[13] Anon., *Scarborough the Queen of Watering Places* (1914), 1 / SCB.

[14] Colin Campbell, *The Romantic Ethic and the Spirit of Modern Consumerism* (York: Alcuin Academics, 2005), 77. The difference between traditional pleasures and modern pleasure is also stressed in Kaj Ilmonen, 'Want, Need and Commodity', in Kaj Ilmonen, Pekka Sulkunen, Keijo Rahkonen, Jukka Gronow, Arto Noro, and Alan Warde (eds), *A Social and Economic Theory of Consumption* (London: Palgrave Macmillan, 2010).

[15] Campbell, *The Romantic Ethic and the Spirit of Modern Consumerism*, 87.

[16] Anon., *Scarborough the Queen of Watering Places* (1914), 1 / SCB. My emphasis.

[17] *The Quiver* 50(8), June 1915, 691 / PQHN.

use of realistic and vivid rhetoric, advertisers present the object of consumption 'in a way that will stand out in the flow of perceptions and memory'.[18] Indeed, contemporaneous resort publicity explicitly promoted the ways in which the emotional experience of the holiday could transcend the duration of the stay at the resort, encompassing both expectations before the visit and the memories that might linger thereafter. A guide to Blackpool noted that 'Blackpool is a perfect haven of hopes, for before its life-giving air, disease of all kinds seems to vanish instinctively.'[19] Likewise, a guide to Skegness mentioned the following:

> To hundreds of thousands of English men, women and children the name of Skegness stands for all that goes to make a perfect and delightful summer holiday ... One goes to it with delightful anticipations, finds in it the realization of one's brightest prospects, and comes away from it regretfully, but with a mind well stored with happy memories.[20]

The emotional characterization of the holiday resort also had gendered overtones. With the shift in the objective of holidaymaking from health to emotional well-being, the role of the female consumer changed as well. Resorts' advertising did not exclusively address men, and in some cases targeted women specifically.[21] One reason for this derives from the inherent logic of advertising, which is meant to appeal to all consumers.[22] A second reason can be found in the ambivalent image of the nervous woman, especially in the context of watering places. On the one hand, if women were prone to nervous disorders, then it was only logical that they would seek out a cure. On the other hand, as Thomas Richards shows in *The Commodity Culture of Victorian England*, female nervousness was being incorporated into the emerging psychology of advertising as an emotional motivation for consumption.[23] As women came to be targeted by experts who subscribed to a gendered view of both nervous disorders and consumer behaviour, some medical conditions were being both exploited and treated by the same industry.

Additionally, in light of the Victorian perception of women as the embodiment of emotionality and irrationality, the idea that holidaymaking was a practice geared towards changing a person's emotions had an interesting impact on women's agency. The emotionalization of the holiday neutralized gender

[18] Eva Illouz, 'Emotions, Imagination and Consumption: A New Research Agenda', *Journal of Consumer Culture* 9, no. 3 (2009): 377–413, here 406.

[19] *Blackpool Pictorial Guide* (1904).

[20] Anon., *Skegness Is So Bracing* (1913), 1 / BL/GRC/YK.1992.a.5634.

[21] Harrogate, for example, published in *The Lady*'s Newspaper and *The Gentlewoman* / NYCRO DC/HRG/BPC 1911–1912; Wells, Baths, Establishment and Advertising Committee, 2 August 1900 / NYCRO DC/RIC II 1/5/9 1896–1900.

[22] Illouz, *Consuming the Romantic Utopia*, 86.

[23] Thomas Richards, *The Commodity Culture of Victorian England: Advertising and Spectacle, 1851–1914* (Stanford: Stanford University Press, 1991), particularly pp. 205–48.

differences to some extent by integrating men into the 'female sphere' of emotions, but it also allowed for a new appraisal of 'female knowledge'.[24] In the emotional realm, women could be the experts and not just the subjects of (male) knowledge systems. This is not to say that actual women were suddenly acknowledged as experts on holidaymaking or emotional well-being, but when describing a resort in 1910, a man could write that 'you should hear also from one of the ladies who have stayed there'.[25] What could a woman add to the official guide books? The writer explained that while '[y]our Guide Book will tell you all about its [the resort's] History and Climate . . . a Lady's impressions of a place are more to be trusted than those of a healthy Man, if we wish to know how enjoyment, peace of mind and ease of body, may be sought and found'. Therefore, the writer asked readers to allow him to 'briefly from the point of view of the "weaker vessels" repeat what I have heard from them'.[26] This allusion to 'female expertise' regarding the experience of holidaymaking, in addition to the usual recommendation letter from a physician that was still inserted in some guides, is a further aspect of the gradual emotionalization of the holiday industry. A year earlier Arthur Laycock's chose to have Sally, a female protagonist as the narrator of the story *Fun on the Pleasure Beach wi' Sally an' Sam, Jimmy an' Jane Ann.*[27] In the following chapter I will suggest that popular seaside stories, such as *Fun on the Pleasure Beach*, were an important source of knowledge for holidaymakers, informing them on the norms, practices, and emotional experiences at the resort. Especially in the case of this story, which is replete with descriptions of feelings and emotional expressions, the choice of a female narrator seems not to be coincidental.

Fun: doctor approved!

The new category of emotion gradually altered the relation between health and amusement, as the legitimized pursuit of emotional well-being opened space for people to enjoy 'healthy gaiety' and 'harmless fun' in a 'soberly gay' resort.[28] At the same time, the positive emotional experiences stimulated by the resort amusements were now being compared to the curative powers of nature: '[t]he piers and

[24] Josephine Kane points to a similar trend in the context of the Edwardian amusement park, and she quotes an American amusement park manager suggesting that 'women are certainly more enterprising than men in collecting new sensations', 'The Taste for 'Thrills', *World's Fair*, 17 August 1912, in Josephine Kane, 'Edwardian Amusement Parks: The Pleasure Garden Reborn?', in Jonathan Conlin (ed.), *The Pleasure Garden, from Vauxhall to Coney Island* (Philadelphia: University of Pennsylvania Press, 2012), 240.

[25] Preface to Mate and Riddle, *Bournemouth: 1810–1910* (1910), i.

[26] Ibid.

[27] Laycock, *Fun on the Pleasure Beach* (1909).

[28] Anon., *Skegness Is So Bracing* (1913), 2 / BL/GRC/YK.1992.a.5634; from a guide to Blackpool, quoted in Webb, 'Bakhtin at the Seaside', 126; from a 1913 guide to a southeast Devon resort quoted in Beckerson and Walton, 'Marketing the Intangible at British Resorts', 59.

promenades and the palaces of pleasure are thronged by a happy holiday crowd, whose fun and frolic are as free, fresh and healthy as the westerly winds of the Lancashire coast'.[29] Once transformed into a matter of health, the emotional discourse shaped the moral framework of holidaymaking by lending scientific authorization to frolic at holiday resorts.

As we can now see more clearly, the emotionalization of holidaymaking did not challenge its therapeutic function; quite the contrary is true. This is evident in the enduring involvement of medical professionals in the industry. In fact, in addition to the ongoing cooperation between medical institutions and the corporations, the pathologization of emotional experiences enabled medical knowledge to penetrate new areas in the sphere of production. Because the pathological condition was defined as a general emotional syndrome and the holiday as an emotional alternation, physicians were concerned not only with the technicalities of the therapy, such as the temperature of the waters or their chemical components, but also with the structure and atmosphere of the holiday as a whole. This holistic approach paved the way for physicians to become involved in all aspects of holidaymaking and the resort. They had to ensure that 'what should be a rest from all worry, a pleasurable, wholesome, and welcome change' would not become 'a matter of drudgery and anxiety, combined with over-crowding, over-exertion, badly cooked meals and general discomfort'.[30]

'Of course, many go to Harrogate who do not need the cure', stated a writer for the *Daily Mail* in 1914. 'They go', he explained, 'because it is an exceedingly pleasant place', adding: 'I do not suppose that anywhere in England is there a finer collection of hotels, and a really good hotel does make a great deal of difference to one's enjoyment of a holiday.'[31] Prima facie, if the visitors 'do not need the cure', then the resort, practices, and people discussed in the text are of little interest to the medical doctor. However, this changes if 'one's enjoyment' is conceived of in therapeutic terms. Indeed, some doctors had been ascribing therapeutic properties to ambience, resort layout, and the like decades before this notion was expressed by resort advertisers and vacationers. Hence, it should not come as a surprise that, when describing the ideal features of the health resort in 1890, MD Burney Yeo also mentioned the '[d]ouble windows, adequate means of heating rooms and passages, carpets, arcades, and loggias for exercise in bad weather, indoor amusements, such as billiard-tables and pianos', further noting that '*all these the doctor recommending* a novel and extreme treatment should be able to guarantee'.[32]

[29] Anon., *A New Pictorial and Descriptive Guide to Blackpool, Lytham, St. Anne's, Fleetwood, and The Fylde* (1902), 3.

[30] Anon., 'Editorials and Notes – Summer Holidays', *The Journal of Balneology and Climatology* 1, no. 3 (1897): 249–51, here 250.

[31] Thorfe Lee, 'Harrogate – A Home Spa', *The Daily Mail*, 22 August 1912 / NYCRO DC/HRG/BPC 1911–1912.

[32] Yeo, *Climate and Health Resorts* (1890), 248. My emphasis.

Yeo was not the only doctor interested in the effects of resort accommodations on the holiday experience. MD George Poore dedicated a whole article in the *Journal of Balneology and Climatology* to the medical aspects of architecture in health resorts, arguing that:

> Hotels, as a rule, are much too high... The lifts and staircases are often without independent ventilation, and each floor ventilates into the floor above and the floor below. The corridors are often close and dark, and there is seldom any adequate provision for flushing them with fresh air. *These huge hotels, where the guests are literally piled in heaps, are most unwholesome. They are the product of cities, and are totally unsuited for climatic health resorts.*[33]

Hotels' architecture was not the only seemingly secondary aspect of vacationing to garner medical interest; the train journey to holiday destinations was scrutinized too. Since a holiday that 'means hurry, overcrowding in railway carriages and lodgings, continual stint, and harassing anxiety as to the expenses incurred, is likely to do more harm than good', doctors advised the railway companies on how to ensure a fatigue-free and healthful journey for holidaymakers.[34] As 'the discomforts and inconveniences are so great consequent upon everybody going away at the same time that not a few have given up the idea of a holiday in absolute despair', holidays needed to be distributed evenly throughout the year, and the train companies were asked to adjust the frequency of the trains and the price of the ride in different seasons.[35]

The letter written by 'A London Physician' touched on yet another dimension of the holiday that others commented on later: A central part of his letter was dedicated to what he called the 'False Economy' of the Harrogate spa, which referred primarily to the custom of tipping the bathers. One of his critiques was that ensuring better and more stable wages for the workers would improve the cure, to which he added the moral argument that tipping is socially unjust. This second point is more interesting for our discussion. The Londoner explained that as the price of a stay in the resort and fees for the treatments and amenities are a substantial expense for many of the visitors, they should not be 'expected to tip the attendant heavily'.[36] In effect, the doctor was advising the resort on how to structure its prices and methods of payment. In a way, this should be seen as an

[33] George V. Poore, 'The Climate of the Dwelling House', *The Journal of Balneology and Climatology* 1, no. 4 (1897): 299–305, here 303. My emphasis.

[34] Anon., 'Editorials and Notes – Summer Holidays' (1897), 249.

[35] Ibid. See also Samuel Hyde, 'Railways and British Health Resorts', *The Journal of Balneology and Climatology* 1, no. 3 (1897): 176–80, 180; Anon., 'The Holiday Huddle', *The Lancet* 2, no. 4534 (1910): 246.

[36] Anon., 'Wake up, Harrogate! A Visiting Doctor's Plain Criticisms', *The Yorkshire Evening Post*, 29 September 1911, 6 / NYCRO DC/HRG/BPC 1911–1912.

early expression of what would become the 'all inclusive' system a few decades later, a system based on the assumption that paying for services during the holiday harms the emotional experience of being 'off work'.[37]

Producing emotional experiences: challenges and contradictions

Echoing the familiar medical ethos, a 1914 GNR booklet titled 'A Few Suggestions for Your Holiday' stated: 'Enjoy the advantages derivable from the force of contrast and the charm of change.'[38] The 'holiday-as-change' ideology promoted an understanding of holidaymaking as a therapeutic technique aimed at providing people with an emotional experience that was deliberately defined as standing in contrast to the emotional conditions of their day-to-day routines. This coincides with John Urry and Ning Wang's conception of tourism as an essentially relational and more specifically antithetical experience that is a cultural response to modern everyday life.[39] However, as I have argued in the introduction, the production of such a contrasting experience is inherently unstable, given that the epistemology of holidaymaking is dependent on multiple and ever-changing perceptions of what causes the negative emotional state that it sets out to relieve, not to mention the shifting meanings of emotions. The act of producing contrast is contingent on the definition of an object from which to differ; but if that initial object is not fixed, the act of producing contrast requires constant updating, both in terms of the knowledge that informs it and the practices that constitute it. In a cultural context composed of multiple and contested views and beliefs, such a process is prone to be fragile and full of contradictions.

Indeed, the production of emotional experiences is by no means a neat process through which an isolated emotional state is formed and experienced in a coherent manner. In contrast to the tendency of 'emotionological' studies to focus on the role of institutions in implementing emotional standards in a 'top-down' fashion, I use the concept of 'emotional economy' also as a way to account for the complexity and ambivalence of processes of creating and circulating emotion knowledge. To that end, I approach the resort industry's production of emotional experiences by analysing the challenges it faced due to the inherent paradoxes of modernity and the ways they were interpreted by different classes and individuals.

[37] Benger Alaluf, 'Exploring the Production of Relaxation in Club Med Seaside Resorts', 44–5.

[38] Anon., 'A Few Suggestions for Your Holiday' (GNR, 1914) / NYCRO DC/HRG/BPC 1914.

[39] Urry, *The Tourist Gaze*; Urry, 'The Place of Emotions Within Place'; Wang, *Tourism and Modernity*.

Capitalizing space and emotions

The previous chapters described a long process of donning upon two spaces the power to evoke emotions antithetical to those of everyday life. First, the problem of overwork was spatially characterized, in that the cause of the negative emotional state was attributed to the modern workplace and the urban environment. At the same time, nature was being constructed as the opposite of urban space, as an emotionally positive space with curative potential. Indeed, resort publicity deliberately referred to the city/nature juxtaposition, explaining that the practice of holidaymaking began 'when London was hidden under a dark and poisonous veil', while visitors at the resort 'were enjoying the bright sunshine, seated in the open air'.[40] It stressed that especially '[i]n these days of over-crowding in cities and towns', the resort's pure air and gardens were the perfect way to 'escape the desolating blight of smoke, dirt and ugliness that has been considered the necessary condition of progress'.[41] And it explicitly claimed that the climate of specific resorts would benefit particularly those who have 'been labouring in the smoky, depressing atmosphere of an industrial centre' and 'jaded workers from busy towns and cities'.[42]

A similar narrative appeared in contemporary literature, for example in 'The Cockney's Paradise (a lay of Margate)', one of Walter Yeldham's[43] poems:

> Are you worn with being busy?
> With the city turmoil dizzy?
> Are you yearning for a pick-me-up
> For jaded flesh and bone?
> Then throw engagements over,
> Take a London, Chat., and Do'ver,
> And, in Margate's bracing breezes,
> Feast your lungs upon ozone.[44]

However, the romantic opposition of city and nature was not without its problems. As one guide to Blackpool noted: 'The glory of the town, and its exceeding

[40] Cole, ed., *Hastings and St. Leonards-on-Sea as a Health and Pleasure Resort* (1884), 25.
[41] Anon., 'Health Resorts: Harrogate', *The Gentleman's Journal*, 1 October 1904, 9266 / HRG/H/AAA.
[42] Anon., *Scarborough the Queen of Watering Places* (1914), 1–2 / SCB; Anon., *Skegness Is So Bracing* (1913), 2 / BL/GRC/YK.1992.a.5634.
[43] Writing under the pseudonym Aliph Cheem.
[44] Aliph Cheem [Walter Yeldham], *Lays of the Sea-Side: A Rhythmical Rendering of Sea-Side Stories and Incidents* (London: Edward Bumpus, 1887), 6.

fame, are the work of the capitalist, not of Nature.'[45] A visit to a holiday resort was far from being an excursion into wild nature; or as some expressed more bluntly: 'seaside places are artificial'.[46] As most notable argued by Peter Borsay, from the outset, resorts were perceived both as 'town' and 'country' and identified with aesthetics and values of both uncultivated nature and urban culture.[47] In this way, they offered the enjoyment of the curative qualities and ideals of nature in a sanitized, cultured locale. Figures 4.1–4.3 illustrate nicely the hybrid character of the resort town: Notice the well-gardened grounds of The Stray in Harrogate (4.1), with its man-made pathways; the massive sea defences, the buildings, and the iconic tower and wheel by Blackpool's sea shore (4.2); the hotel buildings and the cliff lift at Scarborough meeting the sands (4.3).

Much of the work of the corporations was to manage and control nature in order to ensure the safety of the visitors as well as 'the standard of comfort and

Figure 4.1 The Stray, Harrogate, *c.*1895

[45] Anon., *A New Pictorial and Descriptive Guide to Blackpool, Lytham, St. Anne's, Fleetwood, and The Fylde* (1902), x.

[46] Anon., 'Women's Column', *The Cambridge Independent Press*, 3 July 1903, 6 / BNA.

[47] On the ambivalent nature of resorts in relation to the urban/nature divide and their function as both centres of commerce and industry as well as of leisure and recreation, see Borsay, 'Visualising the Seaside, c. 1750–1914'; Borsay, 'Town or Country? British Spas and the Urban–Rural Interface'; Walton, 'Spas in History, Spas and History', 11; Borsay and Walton (eds), *Resorts and Ports*; On the 'Leisure Town', see McInnes, 'The Emergence of a Leisure Town'; Borsay, 'The Emergence of a Leisure Town: Or an Urban Renaissance?'.

Figure 4.2 View from North Pier, Blackpool, *c.*1895

Figure 4.3 The children's corner, Scarborough, *c.*1895

convenience'.[48] Hence, in Hastings, for example, while in the 1870s the Turkish Baths were open to the sea, in the 1880s 'the sea has been kept out by massive walls' in order to protect the visitors from the 'great violence' of the waves.[49] Likewise, along the parade, the corporation installed 'comfortable seats... sheltered from the sun, and protected from the wind and rain by glass'.[50] Blackpool's baths were 'kept at a temperature sufficiently warm to render it pleasant and safe for the most delicate bathers'.[51] In Scarborough, where 'beaches are noted for the smoothness and fine quality of their sands, and their freedom from stones and other impediments to comfortable bathing', the 'Pleasure Boats and Bathing Committee' was asked in a letter of complaint to remove the rocks that were harming 'the condition of the bathing ground in the north bay'.[52] The problem of stones on the beaches appears in other accounts and is brilliantly illustrated in Edwin Wooton's words:

> On the next morning, you sample the bathing. Wherever you pitch your clothes, the stones appear to be much of the same hardness, and those below water to be as disagreeable to kick against... All corporations of rising watering-places are in league with the shoemakers. The former cover the roads with stones, having the size of small cocoanuts, and leave the wearing down process to be done by visitors' boots. The stones return the good endeavour; a steam roller is no longer needed, the shoe trade flourishes, but the visitor mourns.[53]

The demand for smooth sands is particularly interesting, considering that the impetus for bathing was changing towards the end of the century from physical cure to sensorial and emotional pleasure. While bathing as a part of the 'water cure' was traditionally associated with individually experienced pain and despair, once bathing turned into a family diversion, comfort became a legitimate want.[54] Thus, around the turn of the century, many guides described the safety and enjoyment of bathing at a particular resort more than they did its medicinal benefits. This is nicely exemplified in an 1898 Blackpool guide:

> For sea-bathing and boating there is no more delightful and, what is of greater importance, no safer place anywhere... The crowds are merrily disporting themselves enjoying the invigorating sea bath with perfect safety. Visitors enjoy these

[48] Mate and Riddle, *Bournemouth: 1810–1910* (1910), 134–5.

[49] Cole, ed., *Hastings and St. Leonards-on-Sea as a Health and Pleasure Resort* (1884), 5.

[50] Ibid. 6. [51] Anon., *Illustrated Guide to Blackpool and District* (1891), 17.

[52] MacDonald, *An Illustrated and Descriptive Account of Scarborough* (1896), 1–2 / SCB, 2; Scarborough Corporation Minutes, Pleasure Boats and Bathing Committee 1892–, 21 June 1898 / NYCRO DC/SCB II 1/11/2.

[53] Edwin Wooton, 'Where Will You Spend Your Holiday?' *London Society*, August 1898, 116 / PQHN.

[54] Walton, *The English Seaside Resort*, 182.

> pleasures the more by reason of the absence of the irritating shingle and stones and absolute freedom from dangerous currents.[55]

The built environment at the holiday resort also did not differ much from that of the city. In many senses, the holiday resort, while being located 'in nature' and promoted because of this, was not all that different from the city.[56] Interestingly, much resort publicity made this into a selling point, listing it alongside the natural conditions and their contrast to the urban malaise. Skegness was described as 'a thoroughly up-to-date town with a grand sea front, wide well-lighted streets, many imposing buildings, large and beautiful public gardens, first-class hotels and boarding establishments';[57] Harrogate was promoted as 'a place of quiet refinement, with the moors creeping into the town', and in the same breath for its hotels and cuisine, which were 'equal to anything in London'.[58] A Southsea guide mentioned that the town's roads 'are well laid out, paved, and drained', and that '[t]he business streets are broad and well lighted; they contain many large and imposing premises; and the trading thoroughfare, which form charming promenades, present in the season a gay, many-coloured, and busy appearance'.[59] An anonymous commentator actually observed that in the English resort, '[n]either in the main streets nor in the complicated tributaries are there any distinguishing signs of a watering-place . . . excursionists stare into the shop windows, and crowd the sidewalks, and for all that is obvious to the contrary, we might be in some country town on market-day'.[60]

Whereas the medical and the political conceptualizations of holidaymaking mostly emphasized the negative aspects of the urban experience, the ambivalent relation of the industry to urban space followed out of a parallel discourse that should not be underestimated, namely the ideology of modernity and progress and its emotional effects. Alongside nature, the modernity of a resort's infrastructure and amenities was a recurrent motive in travel guides, as expressed in references to their advanced technological features and 'the adaptation of modern science to the needs of life'.[61] Hence, Skegness was described as 'a bright, well-built and perfectly planned Garden City by the Sea', and it made sense to mention in its holiday guide that the local authority 'provided the town with a modern and

[55] Anon., *Blackpool the Unrivalled Sea Side Resort* (1898), 35.

[56] Borsay, 'Town or Country? British Spas and the Urban–Rural Interface'; Walton, 'Spas in History, Spas and History', 11; Walton, *The English Seaside Resort*, 167.

[57] Anon., *Skegness Is So Bracing* (1913), 2 / BL/GRC/YK.1992.a.5634.

[58] Anon., 'Harrogate Is the Most Wonderful of Spas', *Reuter's Bulletin*, Port Said, n.d. / NYCRO DC/HRG/BPC 1914.

[59] Anon., *Sunny Southsea: The Queen of Seaside Health Resorts* (Portsmouth: Holbrook & Son, 1896), 1.

[60] Anon., 'The English at the Seaside', *William H. Harper's New Monthly Magazine* 63 (1881): 484–98, here 490 / PQHN.

[61] Anon., *Sunny Southsea: The Queen of Seaside Health Resorts* (1896), 1.

perfected scientific system of drainage and sanitation, and with a pure and abundant water supply'.[62] Similarly, a guide to Scarborough noted that while the spa 'is by no means a modern institution' and 'has a history of at least three centuries', 'it has been repeatedly modernized so as to meet the changing fashions of the hour'.[63] Every visitor to Bournemouth was said to be 'struck with its appearance of modernity—with all the advantage which that implies', and 'equally impressed with its "old-world" surroundings'.[64] Blackpool corporation labelled the resort 'A Pioneer in Electricity'[65] arguing that it was the first watering place to install electrical lighting in its streets and along the parade, and also proudly mentioning the town's electric tram.[66] The central place of means of transportation in popular holiday literature also hints at the impact that technology and modern infrastructure had on vacationers' ideas about what constituted a good holiday.[67]

Nowhere was 'progress' more clearly manifested than in the spaces of amusement and the pleasure palaces.[68] The spread of commercial entertainment from the 1890s onwards shifted the space of recreation from the outdoors to new complexes, which were the material expression of the shift from nature to artificial amusement: pleasure palaces, amusement parks, aquariums, giant wheels, and towers.[69] As Darren Webb observed in the case of Blackpool, these sites symbolized modernity and displayed the mastery and might of industrial progress.[70] No wonder that Blackpool's tower was erected five years after the inauguration of the Eiffel Tower during the *Exposition Universelle* of 1889 and that its Giant Wheel opened to the public three years after the *World's Columbian Exposition* of 1893. The holiday resort was joining the spectacle culture and engaging with its notions of excitement, novelty, and pleasure by bringing into play the most up-to-date mechanized amusements and the feelings they provoked.[71] The accumulation of human-made features that began covering the resorts' landscapes (e.g. hotels, piers, electric lightning, tram-ways) reached its climax in the erection of towers

[62] Anon., *Skegness Is So Bracing* (1913), 2 / BL/GRC/YK.1992.a.5634.

[63] Anon., *Adey's Annual Popular Guide to Scarborough* (1887), 19.

[64] Mate and Riddle, *Bournemouth: 1810–1910* (1910), 271.

[65] Anon., *Blackpool the Unrivalled Sea Side Resort* (1898), 25.

[66] Ibid.

[67] Margaret Beetham, '"Oh! I Do Like to Be Beside the Seaside!": Lancashire Seaside Publications', *Victorian Periodicals Review* 42, no. 1 (2009): 24–36, here 29.

[68] Kane, 'Edwardian Amusement Parks: The Pleasure Garden Reborn?'; Walton, 'The Re-Making of a Popular Resort: Blackpool Tower and the Boom of the 1890s', 200.

[69] In 1896 Blackpool opened its amusement park, being the first amusement park in Britain. Kane, 'Edwardian Amusement Parks: The Pleasure Garden Reborn?', 221; Walton, *The English Seaside Resort*, 165; 167–70; 173–5.

[70] Webb, 'Bakhtin at the Seaside', 129. For an account of the cultural and emotional meanings of the Edwardian amusement park at the seaside as well as in urban centers, see Kane, 'Edwardian Amusement Parks: The Pleasure Garden Reborn?'.

[71] Webb, 'Bakhtin at the Seaside', 133–5; Kane, 'Edwardian Amusement Parks: The Pleasure Garden Reborn?'. Following Kane's emphasis of the fact that British amusement parks promoted a notion and experience of American culture, the evolution of an emotional experience of 'feeling American' in Britain is an interesting direction for future research.

(e.g. Blackpool 1894; New Brighton, 1900) and giant wheels (e.g. Blackpool 1896), which altered the towns' skylines as if deliberately celebrating the glory of steel over nature.[72] Within the structure of Blackpool's tower was the Tower Circus, a space that could be filled with water and drained rapidly, thus subjecting the essence of a watering place to human manipulation in a sensational spectacle in which the main protagonist was not the sea, but rather modern technology.[73] Paradoxically, the holiday resort was marketed both as untainted natural environment and, in Webb's words, as 'the embodiment of progress',[74] through which humanity was able to control nature for mere entertainment.

The emotional mediation of crowds and soundscapes

Along with the ambivalent relation to nature and modernity, two other similarly contradiction-fraught themes appear in the resort literature and in visitors' reviews: the crowds and their sounds.[75] John Urry distinguishes between two kinds of 'touristic gaze': the 'romantic' gaze, which focuses on solitude in unspoiled nature, and the 'collective' gaze, in which the presence of other people is an intrinsic feature of the place and is vital for its success.[76] Yet, the attribution of one type of gaze or another to a specific space is the outcome of a process that is historically contingent and socially contested. The popularization of the late-nineteenth-century British seaside, as Urry himself rightly notes, was an important moment in this process.[77] Similarly significant to Urry's object of analysis, the gaze, are sound and its cultural framings.[78] Considering the wide-ranging Victorian discourse on the sensory and nervous experience of intense urbanization, it is understandable why overcrowding and noise would pose issues for a space marketed as being antithetical to city life.[79] I suggest that contemporaneous actors' emotional interpretation of people and sounds was based on prior social

[72] Walton, 'The Re-making of a Popular Resort: Blackpool Tower and the Boom of the 1890s', 202.
[73] Webb, 'Bakhtin at the Seaside', 133. [74] Ibid. 124.
[75] Borsay and Walton, 'Introduction: The Resort–Port Relationship', 8–9.
[76] Urry, *The Tourist Gaze*, 43–4; John Urry, 'The Tourist Gaze "Revisited"', *American Behavioral Scientist* 36, no. 2 (1992): 172–86, here 173. See also Urry, 'The Place of Emotions within Place', 78–9.
[77] Urry, *The Tourist Gaze*, 18, 29.
[78] See also Cross and Walton, *The Playful Crowd*, 81. On the place of sound in history see Peter Bailey, 'Breaking the Sound Barrier: A Historian Listens to Noise', *Body & Society* 2, no. 2 (1996): 49–66; Daniel Morat (ed.), *Sounds of Modern History: Auditory Cultures in 19th-and 20th-Century Europe* (New York: Berghahn Books, 2014); Mark M. Smith (ed.), *Hearing History: A Reader* (Athens: University of Georgia Press, 2004).
[79] See for instance Wendy Parkins, 'Trust Your Senses? An Introduction to the Victorian Sensorium', *Australasian Journal of Victorian Studies* 14, no. 2 (2010): 1–7; Jerry White, *London in the Nineteenth Century: 'A Human Awful Wonder of God'* (London: Vintage Books, 2013); Simmel, 'The Metropolis and Mental Life'; Bailey, 'A Historian Listens to Noise', 57–8. On earlier accounts of the city crowd see Ian Munro, *The Figure of the Crowd in Early Modern London: The City and Its Double* (New York: Palgrave Macmillan, 2005); George F. E. Rudé, *The Crowd in History: A Study of Popular Disturbances in France and England, 1730–1848* (London: Serif, 2005).

dispositions, and that for this reason social marketing chose to directly address it in a strategic manner coded with class distinctions.

The crowded piers and promenades were central components of the resorts and were mentioned in numerous accounts of late-nineteenth-century holidaymaking. 'The volume of sound engendered by the vast crowds is prodigious', mentioned 'Lucy' the author of the Ladies' Column of *The Whitby Gazette*. 'It hangs like a blanket upon one's ears… Thousands of separate noises are projected.'[80] The crowds and their related sounds were also valorized in the publicity literature of popular holiday resorts, where 'the shore is crowded with people engaged in every sort of amusement',[81] and where visitors could be found 'mingling with the merry and ever-changing crowds in the streets and on the promenades'.[82] However, the particular emotional meaning attached to the sounds of the crowd is founded on a specific habitus; in other words, it is an expression of a person's embodying a particular social position. Like with other sensory perceptions, the experience of sounds is a product of cultural knowledge just as much as it mediates the social and functions as marker of social identity, thus supporting processes of social differentiation and classification.[83] Noise was a central signifier of identity in Victorian Britain and thus played a crucial role in ongoing class conflicts, which were interpreted by the bourgeoisie as a struggle between refinement and vulgarity, civilization and barbarism. Claiming 'freedom from noise', the British middle and upper classes attempted to contain and regulate the noise of proletarian urbanization and public leisure that they believed was invading their space.[84] The democratization of holidaymaking transferred the negotiation over legitimate sounds in the public sphere to the resort, as is evident from contemporary depictions of the crowded watering place.

A particularly vivid description is found in William Dean Howells' account of his visit to Blackpool in 1911. Interestingly, his reflection pointed precisely to the different emotional meanings that can be attributed to the experience of the crowd:

> From a chair in the hotel garden it looked like some immeasurable organism, some monster of the geologic prime, never still, but creeping with one side this

[80] Anon., Ladies' Column by 'Lucy', *The Whitby Gazette*, 2 August 1907, 7 / BNA.

[81] Anon., *Seaside Watering Places* (1897), 43.

[82] Anon., *Blackpool Pictorial Guide* (1904).

[83] Ori Schwarz, 'The Sound of Stigmatization: Sonic Habitus, Sonic Styles, and Boundary Work in an Urban Slum', *American Journal of Sociology* 121, no. 1 (2015): 205–42; Tripta Chandola, 'Listening into Others: Moralising the Soundscapes in Delhi', *International Development Planning Review* 34, no. 4 (2012): 391–408; Ian Burkitt, 'Social Relationships and Emotions', *Sociology* 31, no. 1 (1997): 37–55. The political role of the sensorial, affective and aesthetic has received increasing attention in the past two decades, mainly as a critique of Habermasian theories of the politics of the public sphere with its primary focus on disembodied reason. See for example: Ruth Leys, 'The Turn to Affect: A Critique', *Critical Inquiry* 37, no. 3 (2011): 434–72; Lincoln Dahlberg, 'The Habermasian Public Sphere: Taking Difference Seriously?' *Theory and Society* 34, no. 2 (2005): 111–36; Ahmed, 'Affective Economies'.

[84] Bailey, 'A Historian Listens to Noise', 57–60.

> way and the other that. Near by it resolved itself into men, women, and children; farther off it was mere human mass... It was always awful to look upon, but awfulest at high noon, when it had swollen to its hugest, and was imaginably famishing for lunch with the hunger of some consuming insect horde. Possibly I am exaggerating in the impression I am trying to give of the Blackpool crowd. *Doubtless any happy couple, near wed or newly wed, of those that abounded in the mass could prove me grotesquely mistaken, if not wilfully false. They could say that they had the time of their lives that day at Blackpool, and could ask nothing better than to repeat their transport.*[85]

Changes in the seaside soundscape also feature in nostalgic narratives about the popularization of the seaside. 'A dozen years ago', mourned a writer in 1889,

> many of the quiet and quaint seaside nooks which nestle in the north-eastern corner of Norfolk were to a great extent unknown regions to the holiday-maker. To-day the restfulness and repose of these old-world villages attract ever-increasing numbers of pilgrims; and the voice of the tourist is now heard in the land of the fern and the poppy.[86]

Health served in some cases to rationalize refraining from visiting crowded resorts. '[A] nervy man or one who is overworked', claimed MD Musgrove, would do better choosing a 'farmhouse holiday' over a popular resort. 'The farmer and his wife probably never heard of nervous debility, and the sufferer therefore does not get a chance of talking about his health, and that is vastly better for him than' being 'surrounded by numbers of people all anxious to talk about their ailments'.[87]

Aristocratic resorts were not necessarily quieter than those popular among the working classes, but their soundscape carried a different meaning in accordance with the habitus of their visitors. An observer of an ordinary morning at Harrogate's Pump Room wrote in 1912:

> By a quarter to eight I am down at the Pump Room. Hundreds of people are there already, drinking their glasses of water and strolling up and down the pretty crescent gardens, where the band is playing as gaily as if it were high noon.[88]

[85] William D. Howells, 'The Waters of Blackpool', *The North American Review* 194, no. 673 (1911): 872–81, here 875; quoted also in Cross and Walton, The Playful Crowd, 98. My emphasis.

[86] Anon., 'Norfolk Sea-Side Nooks', *Eastern Daily Press*, 17 July 1889. Reprinted in Anon., *Holiday Notes in East Anglia: Being a Selection of Articles Which Have Appeared in the Public Press on the Holiday Resorts in Norfolk, Suffolk, and Essex; to Which is Added Information Respecting Railway and Other Facilities for Intending Visitors* (Stratford: GER Company's Works, 1894), 62.

[87] Musgrove, *Holidays and How to Use Them* (1914), 58. Quoted in: 'Holidays for the Nervy', *Daily News & Leader*, 10 April 1914 / NYCRO DC/HRG/BPC 1914.

[88] Thorfe Lee, 'Harrogate – A Home Spa', *The Daily Mail*, 22 August 1912 / NYCRO DC/HRG/BPC 1911–1912.

A similar picture appeared that same year in an article in *The Gentlewoman*:

> [B]y 7.45 you will be drinking your first morning glass, along with a crowd of others, promenading the pretty gardens, and listening to the merry chatter on all hands, and the sweet strains of a band. There is always plenty of music going at Harrogate, from morning to evening, and the best of it, too.[89]

'The setting of the Spa is perfect', one author wrote of Scarborough, 'and the Promenade, as the strains of the Orchestra mingle with the murmur of the sea and the hum of many voices, is a scene of irresistible fascination.'[90] These last accounts describe an intense soundscape that combined human voices and music, which seems to carry a positive connotation for the writers, especially when compared to accounts of the (quite similar) soundscape of working-class seaside resorts, such as this depiction of Blackpool:

> There are scores of peddlers and public entertainments. The music of bands, barrel-organs, fiddlers, harps, and penny whistles, the screams of the bathers, the cries of showmen, the laughter of children, and the entreaties of boatmen swell together in tremendous discord.[91]

So far, this description does not seem to differ much from the examples above. But the following sentence shifts into a degrading tone: 'From ten in the morning until dark the sands are crowded with vigorous and dishevelled holiday-makers, who push and shout as if their enjoyment depended on the extent of their muscular energy.'[92] Clearly, the issue was not so much sound in itself, but the social meaning attached to it.

The interpretation of crowdedness and noise was also shaped by gender. While it is hard to find sources written by women, some insights can still be deduced from the existing material. A remarkable moment appears in *Confessions of a Coquette, while staying at Scarboro', Whitby, & Bridlington*, signed under the pen name Azucena. The story begins with Azucena telling the reader that: 'The best receipt for being soon a bride / Is stay [sic] in summer at the sweet seaside'.[93] Later on she adds:

[89] Anon., 'The Traveler's Bureau: Harrogate, a Delightful Centre for Holiday and Cure Purposes', *The Gentlewoman*, 12 April 1912, 583 / NYCRO DC/HRG/BPC 1911–1912.

[90] Anon., *Scarborough the Queen of Watering Places* (1914), 5 / SCB.

[91] Anon., 'The English at the Seaside' (1881), 498 / PQHN. [92] Ibid.

[93] Anon., *Confessions of a Coquette, while staying at Scarboro', Whitby, & Bridlington. By Azucena* (Scarborough: Dennis, E. T. W, 1888), 3.

On crowded nights to walk about there's no room,
Girls sitting here like dummies in a showroom;
And by their mannners [sic] show they don't forget
Themselves and their affections are to let.[94]

While it is not clear if the text was indeed written by a woman, this stanza plausibly describes an experience of a woman in a crowded resort. Other descriptions of holidays in Women's Columns also suggest they included a lot of work, burden, unpleasant experiences and transgressions.[95] These experiences were not addressed by male commentators. However, the latter did have something to say about the sound of women engaged in leisurely activities. Referring to female factory workers in London, Walter Besant observed that in their free time 'they laugh and shout as they go so noisily that they must needs be extraordinarily happy'.[96] The children's book *At the Seaside* reveals a different example of sonic regulation along gender lines. In the story, 'the boys used to choose such strange places to fish', and they blamed the girls for their failure:

> Fred Murray said it was all the fault of the girls, for they *would* (original) come and sit on the rocks and chatter. What fish *could* (original) be expected to come and help himself to a worm when so much noise was being made? There was more noise then than ever! After that, Fred always went fishing by himself... Fred would never allow the girls near.[97]

The significance of the crowd and the sounds it made for the production of emotional experiences at the resort was twofold. First, given that sounds evoke emotions, they were one in a series of mechanisms that helped constitute the emotional atmosphere of the resort; in other words, they were a central aspect of the product. Thus, it is only logical that resort publicity addressed sound. Second, being that the resort industry was very concerned about the social class of its visitors, it was accordingly sensitive to the moral framing of sounds. One method was to provide visitors with the feeling that they could choose between noisy crowds and silent solitude. 'The charm of Scarborough', argued a 1914 guide, 'is its

[94] Ibid. 20.

[95] Anon., 'Women's Column', *The Cambridge Independent Press*, 27 August 1909, 6; Anon., Ladies' Column by 'Lucy', *The Whitby Gazette*, 2 August 1907, 7; Anon., 'Women's Column', *The Cambridge Independent Press*, 3 July 1903, 6 / BNA. There is no doubt that more research is needed to achieve a clearer understanding of the experience of women in the public sphere, including the history of gendered violence. See Kerry Segrave, *Beware the Masher: Sexual Harassment in American Public Places, 1880–1930* (Jefferson: McFarland, 2014); Judith R. Walkowitz, 'Going Public: Shopping, Street Harassment, and Streetwalking in Late Victorian London', *Representations* 62 (1998): 1–30; Shani D'Cruze, 'Approaching the History of Rape and Sexual Violence: Notes towards Research', *Women's History Review* 1, no. 3 (1992): 377–97.

[96] Besant, 'The Amusements of the People' (1884), 346.

[97] *Anon., At the Seaside* (London: Ward Lock & Co., 1908).

never-ending variety. If you would have a jolly time with crowds of happy people full of the holiday spirit; if you would have a quiet, restful holiday free from rush and bustle.'[98] Other resorts were presented as supplying 'all kinds of amusements... while on the other side of the jetty chairs are provided, and visitors can enjoy perfect quiet'.[99] The beaches of Sussex were promoted as being '"far from the madding crowd", where many a weary brain would gladly rest night by night, and where laughter-loving little urchins could play to their heart's content the whole day long, with no one to say them nay'.[100] A guide to Skegness insisted that although it was a popular resort, '[i]t must not be imagined, however... that Skegness is a place where quietly disposed people cannot enjoy peace and rest when they need them'. Rather, 'there are miles of smooth sands and sheltering sand-dunes where almost the only sounds that are heard are the whispering of the wind in the marram-grass, the crying of the sea-bird and the soothing murmering [sic] of the waves along the shore'.[101]

An additional effect of the noisy/quiet distinction on the holiday industry can be seen in the ways it was used by small, remote resorts in their competition with the larger, more popular watering places. The demand for quiet could be employed to attract 'persons who dislike the somewhat boisterous gaiety of the conventional popular Watering place',[102] or 'those who desire rest pure and simple, without any of those excitements which are to be found at better known watering places',[103] and wish to 'rest and revel in these pleasant nooks, far from the madding crowd, far from nigger minstrel atrocities or fashionable "crawls on the pier"'.[104] In these locations, 'no sound breaks the stillness save the washing of the waves at the base of the cliff, and the rapturous song of the skylark overhead'.[105] It was said that in the quiet resorts

> [t]here is none of the London noise, none of the gaiety of Brighton or Yarmouth, there is no pier, and not even a bathing-machine; but there is a balmy atmosphere, there is peace and quiet, and a sense of rest that is most welcome and beneficial to the mind wearied with brain-work.[106]

[98] Anon., *Scarborough the Queen of Watering Places* (1914), 1–2 / SCB.
[99] Anon., *Seaside Watering Places* (1897), 71.
[100] Blaker, *Burgess Hill as a Health Resort* (1883), 9.
[101] Anon., *Skegness Is So Bracing* (1913), 7 / BL/GRC/YK.1992.a.5634.
[102] Anon., 'Where Shall We Go?' *Household Words*, 10 July 1893. Reprinted in *Holiday Notes in East Anglia* (1894), 48.
[103] Anon., 'Undiscovered Norfolk Watering Places', *The South Eastern Herald*, 29 August 1890. Reprinted in *Holiday Notes in East Anglia* (1894), 60.
[104] Anon., 'Norfolk Sea-Side Nooks', *Eastern Daily Press*, 17 July 1889. Reprinted in *Holiday Notes in East Anglia* (1894), 63.
[105] Anon., 'Where Shall We Go?' *Household Words*, 10 July 1893. Reprinted in *Holiday Notes in East Anglia* (1894), 49–50.
[106] Anon., 'Undiscovered Norfolk Watering Places', *The South Eastern Herald*, 29 August 1890. Reprinted in *Holiday Notes in East Anglia* (1894), 60.

Another way in which the holiday industry addressed the morally-charged issue of noise was to neutralize it through emotional language: sometimes, the same elements that were criticized in reviews of popular watering places were re-articulated in the resort publicity in positive emotional terms. Consider the following description published in the Blackpool corporation guide of 1898:

> Sounds of pleasure and merry-making greet the ear as we mix with the blithe and bustling crowd. The Tower looks down from the blue vault of space, with tiny looking creatures disporting themselves on its dizzy eminence. The Gigantic Wheel revolves with its happy human cargoes—laughing, shouting, singing as they go . . . The Piers are covered with living streams of people, and electric trams glide to and fro, each with its complement of sight-seers. On the three miles of Promenade—on the very brink of the glorious waters, thousands of all conditions of people drink in the phosphorous and ozone of the life-giving breezes . . . All around is the music of joy's great orchestra, harmonized with the measured cadence of the sounding sea. A bright, breezy morning at Blackpool lives in the memory![107]

In contrast to the previous examples, here the corporation was not trying to avoid the 'problem' by shoving it aside to a designated space or a particular season, but instead sought to break the identification of sounds as noise and of people as noisy by applying an emotional filter: 'sounds of pleasure and merry-making' are not interpreted as rudeness, but as 'a greeting to the ear'; people 'laughing, shouting, singing' are not 'vigorous and dishevelled holiday-makers', or a 'monstrous human mass', but 'happy humans'. A similar emotional interpretation was applied to 'the crowd', which was referred to in the guide to Scarborough quoted above as 'happy people full of the holiday spirit'.[108] This articulation reveals the role that the holidaymaker was to take in their own vacation: since each person was positioned both as consumer and as producer of the emotional experience, 'the happiness of the place seems to grow with the density of the throng'.[109] Once crowdedness and noise were harnessed for the production of the emotional experience, they could be liberated from their negative urban connotations and become representative aspects of vacationing, or in the words of 'Lucy: 'One discovers the foundation of popular enjoyment to be noise'.[110]

[107] Anon., *Blackpool the Unrivalled Sea Side Resort* (1898), 25.
[108] Anon., *Scarborough the Queen of Watering Places* (1914), 1–2 / SCB.
[109] Anon., *A New Pictorial and Descriptive Guide to Blackpool, Lytham, St. Anne's, Fleetwood, and The Fylde* (1902), 16.
[110] Anon., Ladies' Column by 'Lucy', *The Whitby Gazette*, 2 August 1907, 7 / BNA.

Rest till you drop? The threat of boredom

The ambivalent relationships between nature/modernity, crowds/solitude, and noise/quietness reflect the paradoxical nature of the production of emotional experiences. The analogous tension in the realm of entertainment was that between rest and boredom, tranquillity and dullness. As is well known, this tension was observed and analysed by contemporary intellectuals in the context of the emerging urban mass consumption culture in Europe and the United States, and has continued to draw much attention since.[111] Many accounts interpreted popular entertainment as providing a distraction from the feeling of alienation caused by modern urban life, distractions that usually involved intense sensations of excitement and thrill stimulated by modern industrial technology. However, most accounts also pointed to the inherent insufficiency of the distractions, as they themselves were, so critics said, alienated and false in essence.[112] This lack of satisfaction, it was argued, caused consumers to engage in an endless pursuit of different, novel attractions, which in turn was alleged to generate a state of ennui, boredom, and, consequently, disappointment.[113]

Holidaymaking is an interesting form of leisure consumption, because its purpose rests on the assumption that it provides a restful break from daily routines. Still, a central promise of resort marketing, especially after the turn of the century, was the commitment to provide visitors with endless entertainment options and an experience that will be 'never dull'.[114] Almost all resorts mentioned that they offered 'many attractions',[115] 'varied entertainments',[116] 'attractions too many for specification',[117] 'much to please, to admire, and to attract',[118] 'much to amuse, interest and instruct',[119] 'plenty to do and plenty to see',[120] and so on. In Skegness, amusements were said to be so many 'that it is impossible for a visitor to

[111] See for instance Simmel, 'The Metropolis and Mental Life'; Maxim Gorky, 'Boredom', *Independent* 63, no. 8 (1907): 310–5; David Frisby, *Fragments of Modernity: Theories of Modernity in the Work of Simmel, Kracauer and Benjamin* (Plymouth: Routledge, 2013); Tom Gunning, 'An Aesthetic of Astonishment: Early Film and the (In)Credulous Spectator', in Philip Simpson, Andrew Utterson, and Karen J. Shepherdson (eds), *Film Theory: Critical Concepts in Media and Cultural Studies Vol. 3* (London: Routledge, 2004), 78–95; Walter Benjamin, *The Arcades Project*, trans. Howard Eiland (Cambridge: Harvard University Press, 1999 [1927–1940]); Williams, *Dream Worlds: Mass Consumption in Late Nineteenth-Century France.*

[112] Gunning, 'An Aesthetic of Astonishment', 114–33.

[113] Ibid. 126; Williams, *Dream Worlds: Mass Consumption in Late Nineteenth-Century France*, 57.

[114] Anon., *Blackpool the Unrivalled Sea Side Resort* (1898), 39. See also Walton, *The English Seaside Resort*, 42.

[115] Anon., 'Scarborough – "Queen of Watering Places"', *Vanity Fair*, 25 June 1914 / NYCRO DC/HRG/BPC 1914.

[116] *Town & Country*, 6 May 1912 / NYCRO DC/HRG/BPC 1911–1912.

[117] MacDonald, *An Illustrated and Descriptive Account of Scarborough* (1896), 1–2 / SCB, 10.

[118] *The Daily Graphic*, 20 July 1912 / NYCRO DC/HRG/BPC 1911–1912.

[119] Anon., *Sunny Southsea: The Queen of Seaside Health Resorts* (1896), 47.

[120] Anon., Scarborough The Queen of Watering Places (1914), 34 / SCB.

exhaust them or ever to spend a dull hour';[121] in Harrogate, '[r]ecreations are far too numerous to be particularized', and they 'succeed each other with marvelous rapidity';[122] and in Blackpool one can stay for 'a month and find a new amusement every day'.[123] This means that the British holiday was, at least in one respect, not all that different from other forms of mass consumption criticized by contemporaries.

In theory, facing such a wide variety of entertainment opportunities might easily induce anxiety rather than relaxation, information overload, and a fear of disappointment.[124] Some descriptions of the assorted recreational facilities seem almost claustrophobic:

> In one room the conjuror plies his magic art; in another the humorous of a high-class variety show delight a packed audience; down below a drawing-room, dramatic entertainment is in progress; up above, here and there, no matter which way we turn, something interesting is going on; and over the happy, light-hearted hum that pervades the whole, float the strains of the waltz of the music.[125]

It should be added that each attraction usually had its own entry fee, meaning that choosing entertainment was also an economic choice in which visitors had to take into consideration their purchasing power and the related opportunity costs. Only after the turn of century did some resorts begin offering a combo-ticket to their pleasure palaces, so that 'on payment of the modest sum of sixpence one can enjoy a 10 hours' continuous programme'.[126] This approach was the progenitor of the 'all inclusive' logic applied in holiday resorts in the second half of the twentieth century, which served to eliminate the potential stress associated with consumerist choice.[127] Furthermore, while early twenty-first century resort publicity emphasizes the legitimacy of the option to choose not to choose and simply 'do nothing',[128] Edwardian holiday culture exerted pressure on the holidaymaker to exhaust all the choices available to her. Thus, for example, after listing the amazing

[121] Anon., *Skegness Is So Bracing* (1913), 1 / BL/GRC/YK.1992.a.5634.

[122] Anon., *My Holiday: Where Shall I Spend It?* (London: Sampson Low, Marston, Searle, and Rivington, 1878), 43.

[123] Anon., 'Manchester at the Seaside: Prosperous Blackpool', *The Manchester Guardian*, 11 July 1896, 4 / PQHN.

[124] Barry Schwartz, *The Paradox of Choice: Why More is Less* (New York: Ecco, 2004); Sheena S. Iyengar and Mark R. Lepper, 'When Choice is Demotivating: Can One Desire Too Much of a Good Thing?' *Journal of Personality and Social Psychology* 79, no. 6 (2000): 995–1006; MacCannell, *The Tourist: A New Theory of the Leisure Class* (Berkeley: University of California Press, 1976), 49; Simmel, 'The Metropolis and Mental Life'.

[125] Anon., *A New Pictorial and Descriptive Guide to Blackpool, Lytham, St. Anne's, Fleetwood, and The Fylde* (1902), 6.

[126] Anon., *Scarborough the Queen of Watering Places* (1914), 34 / SCB.

[127] Benger Alaluf, 'Exploring the Production of Relaxation in Club Med Seaside Resorts', 48–9.

[128] Ibid. 48.

variety of attractions, an official guide to Scarborough warned the visitor that he should not leave any stone unturned 'if he is to enjoy his holiday to its fullest extent, and derive from it the greatest benefit'.[129] These words reflect a mentality of responsibilization that would become a prominent feature of consumer culture in neoliberal societies.[130] In light of the expectation that visitors take advantage of all of the resort's entertainment options, one guide to Blackpool noted that '[a]n especial feature of this town, and which is not possessed in the same degree by any other seaside place of resort, is the centralization of its amusements and places of public interest'. Such centralization was a major relief, 'as no special effort is necessary to get from one place of pleasure to another'.[131]

The sources analysed it this chapter challenge the common view of the holiday as 'escape' or as the 'other'. In reality, the experience of holidaymaking reflected an epistemology related to a matrix of various emotions, dispositions and meanings. Holiday resorts did not supply only change or contrast; they also provided much of the sense of normality, security and familiarity.[132] In addition, and in contrast to Herbert Marcuse's claim that consumer demand is an expression of 'false needs' controlled and manipulated by the production system,[133] the case of the holiday industry reveals a different dynamic. Once emotional experience came to determine the value of the product, the resort had to shape its atmosphere to harmonize with multiple and unstable perceptions of the emotional meaning of modernity, morality, and health. In this sense, holiday consumption was not entirely subordinated to the production sphere. Rather, the value of the holiday was co-produced: its fulfilment, or success, was possible only if it fit consumers' desires, which were in turn related to the wider emotional economy.[134] This complex and unfixed relation between emotions and consumption will be explored in the following chapter.

[129] Anon., *Scarborough the Queen of Watering Places* (1914), 35 / SCB.

[130] Edgar Cabanas, 'Rekindling Individualism, Consuming Emotions: Constructing "Psytizens" in the Age of Happiness', *Culture & Psychology* 22, no. 3 (2016): 467–80; Mike Featherstone, 'The Body in Consumer Culture', in Jessica R. Johnston (ed.), *The American Body in Context: An Anthology* (Wilmington: Scholarly Resources Inc., 2001), 89.

[131] Anon., *Illustrated Guide to Blackpool and District* (1891), 27.

[132] See also Kane, 'Edwardian Amusement Parks: The Pleasure Garden Reborn?', 240–1; Borsay, *A History of Leisure*, 172; Edensor, 'Performing Tourism, Staging Tourism', 61.

[133] Herbert Marcuse, *One-Dimensional Man: Studies in the Ideology of Advanced Industrial Society* (Oxon: Routledge, 2013 [1964]).

[134] Cf. Rafael Ramirez, 'Value Co-production: Intellectual Origins and Implications for Practice and Research', *Strategic Management Journal* 20, no. 1 (1999): 49–65, here 55.

5
Learning to consume emotional experiences

Besides vacationers' numerous accounts of successful holidays, many accounts also explicitly described their failure. These stories of disappointment are often narrated in the plural, implying that it was common knowledge that holidays frequently went wrong. As early as 1871, the author of an article titled 'Holidays' wrote that 'though we have all vowed a good many times that we had rather have stayed at home, and meant to do so in the future', in reality 'we never had any serious intentions of doing anything so silly, and this year we shall take or have taken our holiday as usual'.[1] A few decades later, the author of a Women's Column seemed to be more open to the possibility of dispensing with the holiday altogether, enumerating some of the inconveniences of holidaymaking:

> Many of us have, I dare say, already made up our minds where to go for our summer holiday... but to those who are still weighing the respective merits of Hunstanton and Cormer, Margate and Clacton, may I humbly suggest a solution of the problem—why not stay at home? Think of your experiences at the seaside last year. Your lodgings were small and only cleanish ... You spent nearly all your mornings trying to get served at crowded, stuffy shops... In the afternoon you sat on the beach where hundreds of other mortals were trying to delude themselves into thinking they were tasting perfect bliss. The sands were covered with paper bags, gooseberry skins, straws, buttons, and odds and ends 'enough to fill nine museums'... Do you remember what a long time you had to wait for your bathing machine, and how cross you were, and how the sea was full of dead fish and corks and bits of sticks? Do you remember how you met Mrs. A, the lady whom you have for years been trying to avoid?... Do you remember—but of course you do—all the little incidents of your stay, *and* (original) the bills at the end.[2]

What can we learn from these accounts? Economists, psychologists, and sociologists studied frustration, boredom, and disappointment as features of consumer culture and have asserted that consumption induces these emotions precisely because they increase the desire to consume, revolving in an invisible dynamic

[1] Anon., 'Holidays', *The Bolton Chronicle*, 16 September 1871, 5/BNA.
[2] Anon., 'Women's Column', *The Cambridge Independent Press*, 15 August 1902, 6/BNA.

The emotional economy of holidaymaking: Health, pleasure, and class in Britain, 1870–1918. Yaara Benger Alaluf, Oxford University Press (2021). © Yaara Benger Alaluf. DOI: 10.1093/oso/9780198866152.003.0006

of imagination–frustration–refinement.[3] *Emodities*, however, have to be consumed by an emotional subject, and their ability to generate profit depends on the compatibility of how the product is experienced with existing emotion knowledge.[4] Holidaymakers became emotional subjects and emotional consumers through a learning process.[5] They were active agents in the construction of emotion knowledge, and their perception and evaluation of their experiences at the resort fed back into holiday culture, contributing to its gradual transformation. In order to demonstrate this, rather than simply accounting for the emotions that individuals associated with holidaymaking, I explore their exposure to and construction of available emotion knowledge related to holidaymaking. I approach the emotions of disappointment and frustration, especially, not as merely mechanisms that increase the desire to consume, but as moments that expose the gap between expectation (knowledge) and experience (consumption).[6]

Disappointment and failure

Vacationers referred to a wide combination of experiences that put a damper on their holidays, including 'vacant hours, bad cooking, and dear lodgings', which 'sour the temper, create an abnormal thirst, and derange the stomach'.[7] 'We all go to the seaside', stated a contributor to the *Bolton Chronicle*, 'and a terrible business it is for some of us.'[8] Among the potential inconveniences the author mentioned were troubles and misunderstandings about luggage and accommodations, and delayed trains carrying 'an invading army of excursionists about four hours late at a small town already overcrowded'.[9] Even John Hartley's humorous seaside story *Seets i' Blackpool* describes in detail the trouble of finding affordable accommodations while schlepping one's luggage around town.[10] Advising her readers to avoid holidays at British seaside resorts, the author of *The Cambridge Independent Press*' Women's Column went so far as to argue that '[i]t is a mere push and scramble for board and lodging and elbow room. It makes one disgusted with

[3] Illouz, 'Emotions, Imagination and Consumption'.

[4] Illouz, ed., *Emotions as Commodities.*

[5] In his influential article 'Performing Tourism, Staging Tourism' Tim Edensor emphasizes the argument that tourists are in a constant process of 'becoming' and learning 'how to be a tourist'. This chapter provides an empirical account of the performative nature of holidaymaking, focusing (to a greater extent than Edensor) on the emotional sphere. Edensor, 'Performing Tourism, Staging Tourism'.

[6] For further reading about 'failure' and the history of emotions, see William M. Reddy, 'Against Constructionism: The Historical Ethnography of Emotions', *Current Anthropology* 38, no. 3 (1997): 327–51; Boddice, *The Science of Sympathy*, 21–2; Scheer, 'Are Emotions a Kind of Practice', 211.

[7] Anon., 'Bored at the Sea-Side', *The Observer*, 20 August 1871, 5/PQHN.

[8] Anon., 'Holidays', *The Bolton Chronicle*, 16 September 1871, 5/BNA.

[9] Ibid.

[10] John Hartley, *Seets i' Blackpool, Fleetwood, Lytham, and Southport: As Seen bi Sammywell Grimes an' his Wife Mally on Their Hallidy Trip, wi' a Few Incidents an' Accidents 'at Occurred* (London: W. Nicholson & Sons, 1890).

humanity that is so numerous and so noisy.'[11] Frequent cases of disappointment could turn into anger directed towards guides, advertisements, and 'the asterisked liar, who led you to this asterisked place of fraud, torture and desolation'.[12]

In addition to the frustration caused by the 'very dull grey'[13] reality, and not unlike other spheres of consumption, holidaymaking was associated with an intrinsic anxiety about others' opinions on one's desires and choices. This kind of emotional burden, as was explored by Susan Matt in the case of the United States,[14] also encumbered potential holidaymakers in Great Britain. 'Fashion and self-importance step in', admitted one writer, 'and humble Hornsea succumbs to breezy Bridlington. Smithers is successful in business, and goes there for a month. *We* (original) are also prospering, and must follow suit, otherwise what will Smithers think?'[15] Others were disappointed with their holidays' romantic outcomes. 'Lucy', the author of the Ladies' Column of *The Whitby Gazette*, described the whole process: the notion of the holiday as a pleasant chapter in one's life, the expectations generated by this perception, and the painful disappointment. The author not only described a given experience, but sought to educate readers about the emotions attached to it and how to avoid negative emotions:

> With a not inconsiderable number of people, life goes in chapters. A very pleasant one begins, usually, when they go away for the summer holiday. In the case of the average girl, this chapter unfolds itself to very interested eyes. She expects it to be rather romantic. Sometimes it is, sometimes it is not. Anyway, many a maiden looks upon a seaside holiday as a period crowded with possibilities. Often, however, her fairy fancy pictures impossibilities as possibilities. Then, I fear, is she doomed to disappointment. Don't expect too much of the holiday, girls. Don't put too many stories on your castle in Spain, and remember that the imaginary picture is too highly coloured, the ordinary black and white of the actual happenings may, in contrast, appear to be a very dull grey. This is sound advice, though I know that, as a rule, young folks in holiday-making intent are not prone to like sound advice.[16]

'Nothing spoils one's pleasure so much as having to count sixpences', and indeed, 'the very vulgar consideration of money' was for many a major source of frustration.[17] As a writer observed in 1880, 'the holiday-maker of the present day may with much better reason complain that the multitude of trips and excursions has

[11] Anon., 'Women's Column', *The Cambridge Independent Press*, 3 July 1903, 6/BNA.
[12] Wooton, 'Where Will You Spend Your Holiday?' (1898), 116.
[13] Anon., 'Ladies' Column by "Lucy"', *The Whitby Gazette*, 2 August 1907, 7/BNA.
[14] Matt, *Keeping Up with the Joneses.*
[15] Anon., 'On My Back', *The Hull Packet and East Riding Times*, 25 June 1880/BNA.
[16] Anon., 'Ladies' Column by "Lucy"', *The Whitby Gazette*, 2 August 1907, 7/BNA.
[17] Anon., 'Holidays', *The Bolton Chronicle*, 16 September 1871, 5/BNA.

been the means of robbing him of much statute made importance'.[18] Even visitors to aristocratic Harrogate mentioned that the 'daily fees for the cure, what with waters, baths, massage, douches, etc., are such a considerable item to a large number of the people who come'.[19] The impact of the costs of the holiday was loudly echoed in popular literature, where descriptions of the act of paying and exact descriptions of the prices of a resort's activities appear frequently. In his *Adventures and Sketches at New Brighton*, Alfred Short's alter ego 'One Short' meets a young woman with whom he enjoys the various attractions of the seaside resort. Here is one of the couple's 'adventures':

> We chanced upon a Photographic Artist at his studio door,
> Seeing we were subjects likely, he asked us most politely
> If our photos, he should take, he would'nt charge us any more
> Than sixpence each if plain, or if you'd wish them color'd, sixpence more,
> They will last for evermore.
> Before we could well decide, he inveigled us inside,
> And grouped us side by side, upon his studio floor,
> And in wond'rous little time, produced two photos, fine,
> Frames, he said, are fourpence more, altogether one and four,
> A memento of your visit and your lady's to our pleasant shore,
> If you don't come anymore.[20]

Prices are mentioned so many times in this short passage that they overshadow the main theme of photography as a holiday practice. In Arthur Laycock's story *Fun on the Pleasure Beach wi' Sally an' Sam, Jimmy an' Jane Ann*, too, the costs of activities are incorporated in the narrative, as are remarks such as 'it's cheap enoof' or 'we began emptyin' eawr pockets again'.[21] And Ab-o'-th'-Yate describes Blackpool as a place 'where folk go'n to ut wanted to be quiet, an' do things chep'.[22] Economic expenses seem to have been so significant during the rising popularity of holidaymaking that prices of activities were even mentioned in poems and children's books. *Sally and Mary's Seaside Holiday Pictures* tells the story of the daughters of 'very poor people living in London', sent to spend a holiday at the seaside for the first time in their lives. The fact that the girls come

[18] Anon., 'On My Back', *The Hull Packet and East Riding Times*, 25 June 1880/BNA.

[19] Anon., 'Wake Up, Harrogate! A Visiting Doctor's Plain Criticisms', *The Yorkshire Evening Post*, 29 September 1911, 6/NYCRO DC/HRG/BPC 1911–1912.

[20] One Short [Alfred Short], *Adventures and Sketches at New Brighton by One Short* (Liverpool: A. Short, 1888), 4.

[21] 'It is cheap enough'; 'After filling our stomachs, we began emptying our pockets again'. Laycock, *Fun on the Pleasure Beach* (1909), 4–5, 12.

[22] 'Where folk go on to that wanted to be quiet, and do things cheap'. Ab-o'-th'-Yate [Benjamin Brierley], *Drop't on at Blackpool* (1881), 15.

from a poor family opens the story, and the plot revolves around their tight budget and how it forbids them from taking pleasure in the 'delights' of seaside activities, demoting them to mere observers 'watching more fortunate children enjoying themselves'. One of the sections of the book begins with the words: 'Now for a jolly sail, two hours on the sea for sixpence', recounting how Mary and Sally begged a boatman to give them a discount because they 'have not much money'.[23] Indeed, holidaymakers were surrounded by advertisements placed in train stations; in the covers of guide books, journals, and seaside stories; and even in bathing machines, where 'sometimes the whole of the superficial space is covered with advertisements'.[24]

With ever more working-class people going on holiday, resort publicity started addressing visitors' buying power and economic considerations. Blackpool, for example, was described as a place where 'more fun could be found, for less money, than anywhere else in the world'.[25] Similar statements included: 'anyone can find amusement for a small fee', or 'so joyous and exhilarating—*and* (original) so cheap'.[26] Some reviewers and guides recommended particular resorts where attractions could 'be enjoyed at an exceedingly moderate tariff' or specifically wrote for 'those to whom economy is a vital consideration', 'those requiring a quiet, clean, healthy, and cheap watering-place', or 'the professional man with a young family, not too well endowed', who 'will be surprised at the reasonableness of the charges'.[27]

The centrality of economic considerations and related anxieties were closely connected to the central role of social class in Victorian and Edwardian holiday culture. We have seen that resorts were intensely concerned about the class of their visitors and actively addressed the issue both through publicity and by adjusting facilities to a desired level of 'respectability'. Vacationers' accounts and holiday literature reflected the influence of class on the emotional experience of holidaymaking. Deliberately or not, many narratives often shed light on the social composition of the resort. For instance, in *The Cockney's Paradise (a lay of Margate)*, 'Aliph Cheem' writes:

> Let us up and take a ramble,
> Or, more properly, a scramble,

[23] Anon., *Sally and Mary's Seaside Holiday Pictures* (London: R. Culley, 1907).

[24] Anon., 'The English at the Seaside' (1881), 493/PQHN; see also Beetham, 'Lancashire Seaside Publications', 27; Wells, Baths, Establishment and Entertainments Committee, 14 February 1901/ NYCRO DC/RIC II 1/5/10 1900–1906; Scarborough Corporation Minutes, Parks, Advertising and Entertainments committee 1909–1911, 16 March 1910/NYCRO DC/SCB II 1/13/3; Simmons, 'Railways, Hotels, and Tourism in Great Britain 1839–1914', 217.

[25] *The Morning Post*, 24 August 1887, quoted in Urry, *The Tourist Gaze*, 23.

[26] Anon., *Seaside Watering Places* (1897), 17; Anon., *A New Pictorial and Descriptive Guide to Blackpool, Lytham, St. Anne's, Fleetwood, and The Fylde* (1902), 3.

[27] Anon., 'On My Back', *The Hull Packet and East Riding Times*, 25 June 1880/BNA; Anon., *Seaside Watering Places* (1897), 41, 381, 445.

Through the crowds at Cockney Margate
In the eighth month of the year.
'Tis the climax of the season,
And you'll find, I say with reason,
There are funny things to look at there,
And sunny things to hear.
Though the language mayn't be Attic,
Nor the folk aristocratic.
Though the dresses aren't exactly
Such as Hurlingham displays;
If you're not a stuck-up party,
And can love what's bright and hearty,
With a sniff of brine, at Margate
You may spend some cheery days.[28]

Similar references to the class of the visitors appear in several of his poems, and the concluding song, 'The Seaside of England', starts with the following words:

The Seaside of England! how dear to us all!
Be we rich, be we poor, be we full-grown or small!
Be we well, be we ill, be we married, or single—
We all love a glimpse of the insular shingle.[29]

The seaside might have been as dear to the rich as to the poor, but Cheem's inclusive rhetoric is misleading. The encounter between different classes at the holiday resort ultimately resulted in people constantly identifying and labelling 'the other', thus maintaining and reinforcing established social structures. 'The Brighton crowd', stated a visitor, 'is usually well-behaved under all circumstances, but it is composite; and that noisy and unwholesome creation of modern English life, the London 'Arry, is here among the rest.'[30] Similarly, in the description of her visit to the Yorkshire coast, 'Azucena' mentioned her encounter with 'the cheap tobacco-scented tripper, [w]ith ill-made clothes, and face not fair to see', and later on declared that she 'should not care to dance with young men without money', even if they were 'amusing and funny'.[31]

The encounter between different classes at the holiday resort was an intense emotional experience that stimulated both excitement and fear. Pondering past

[28] Aliph Cheem [Walter Yeldham], *Lays of the Sea-Side* (1887), 6. [29] Ibid. 229.

[30] Anon., 'The English at the Seaside' (1881), 495/PQHN. "Arry' was a derogatory term for members of the lower middle classes in late Victorian Britain based on Edwin J. Milliken's series *The 'Arry Papers*.

[31] Anon., 'Confessions of a Coquette, while Staying at Scarboro', *Whitby, & Bridlington* (1888), 19, 125.

vacations at the beaches of the East Riding of Yorkshire and north Lincolnshire, a middle-class writer—as can be deduced from his use of different languages and acquaintance with continental Europe—described the 'crowds of pleasure-hunters-toilers in the great seething of factories and iron mills of the Midlands—unkempt and unwashed, and with prodigious appetites for enjoyment'. His account reveals mixed feelings towards the rude and dirty 'hardy sons of industry' and their 'happy innocence and glee', and the encounter is described as an attraction in its own right, which 'must be *seen* (original), not described!'[32]

At the opposite end of the class spectrum, these semi-anthropological encounters evoked not only curiosity and astonishment, but also feelings of discomfort and worry. In one of their visits to Blackpool, Ab-o'-th'-Yate and his friends are passed by a carriage of upper-class women, and Ab notes 'th' women gan us a look as if th' next move would be sheautin' for a policeman'.[33] Blackpool offers a particularly interesting perspective on class relations. It is rightly referred to as the exemplary working-class seaside resort, but precisely because of this it was also the holiday resort with the sharpest geographical class segregation.[34] When William Dean Howells announced that he was planning to visit Blackpool, it filled his 'English friends with surprise and pleasure'. He noted that he and his companion 'were, in fact, so high in the social scale' that it was astonishing that they even knew about the place. In fact, they had heard about it from 'one of the most liberal of' their friends, who knew about Howells' 'interest in the popular crowd'. However, once in Blackpool, he acknowledged the diversity of the place that brings together 'masters and men alike'.[35]

Indeed, members of the middle and upper classes did visit Blackpool, and the corporation made an effort to meet their expectations and preserve a steady flow of visitors from the middle classes. This was done by the means of: (a) the constitution of a system of social zoning starting in the 1860s—the middle classes remained mainly on the North Shore, where the pier was segregated from the lower classes by a two-penny toll and the area was protected by special laws against noise and street showmen; and (b) the implementation of a policy of regulations in the town centre, where most amusements and shops were concentrated and social mixing was harder to prevent.[36] These methods of class

[32] Anon., 'On My Back', *The Hull Packet and East Riding Times*, 25 June 1880/BNA.

[33] 'With two women in, sitting as stiff as two queens ... the women gave us a look as if the next move would be shouting for a policeman'. Ab-o'-th'-Yate [Benjamin Brierley], *Drop't on at Blackpool* (1881), 16.

[34] Walton, 'Residential Amenity, Respectable Morality, and the Rise of the Entertainment Industry'; Webb, 'Bakhtin at the Seaside', 125.

[35] Howells, 'The Waters of Blackpool' (1911), 872.

[36] Walton, 'Residential Amenity, Respectable Morality, and the Rise of the Entertainment Industry'; Perkin, 'The "Social Tone" of Victorian Seaside Resorts in the North-West', 180, 188.

segregation were a known fact and were directly referred to in guide books and in popular 'Trip to Blackpool' stories.[37]

It is not hard to find evidence for the emotional impact that this reality had on holidaymakers. During their Blackpool adventure, Ab-o'-th'-Yate and his mates 'dar' go closer to' upper-class holidaymakers at the new pier 'witheaut fear o' bein' towd by looks, of not by words'. Once closer to this 'sort o' folk', Ab-o'-th'-Yate reports his impressions: '[t]heere noses are cocked up, an' legs are swingin' in a hurried walk', they are organized in groups, their tobacco smells stronger, the women are louder, their clothes and hats are different, and in contrast to Ab and his friends, the people on the new pier enjoy their holiday without any 'unyezziness abeaut what it cost 'em'.[38] The adventure on the new pier ends without complications, but Brierley's choice to describe his main character's worries shows how emotionally loaded such encounters could be. Sammywell and his wife Mally, the protagonists of John Hartley's *Seets i' Blackpool*, also experience a kind of culture shock on the pier:

> 'Ha mich is it to goa in?' aw axed. 'Twopence each!' an' aw put daan a fourpny bit an' th' windlass clickt twice, an' we wor on th' North Pier. It wor hard to fancy 'at ther could be soa mich poverty, an' misery o' one sooart an' another I'th' world, when aw lukt at th' craads o' happy lukkin' mortals 'at wor sittin' or walkin' past,—old fowk at lukt as if they nivver had a care, an' young ens 'at lukt too beautiful an' pure to be classed amang 'miserable sinners', an' rosy little rollickin' lads an' lasses, soa full o' frolic, soa free o' care.[39]

In contrast to the story of Ab-o'-th'-Yate, the narrator reveals nothing about what Mally and Sammywell felt before going up to the pier, but Hartley does dwell on Mally's emotions at this particular location, 'the forbidden pier':

> Ivverything lukt strange an' grand to Mally, an' as we sat daan for a bit ov a rest, aw nooaticed 'at her face wore th' same serene smile at aw've seen on it when shoos been sat listenin' to a sarmon throo one ov her favorite texts, an aw knew

[37] Webb, 'Bakhtin at the Seaside', 125.

[38] 'Dare go closer to'; 'without fear of being told by looks, if not by words'; 'sort of folk'; 'their noses are cocked up, and legs are swinging in a hurried walk'; 'uneasiness about what it costs them'. Ab-o'-th'-Yate [Benjamin Brierley], *Drop't on at Blackpool* (1881), 14–15.

[39] '"How much is it to go in?" I asked. "Twopence each!" and we put down a fourpenny bit and the windlass clicked twice, and we were on the North Pier. It was hard to fancy that there could be so much poverty, and misery of all one sort and another in the world, when we looked at the crowds all happy looking mortals that were sitting or walking past, – old folk that look as if they never had a care, and young ones that look too beautiful and pure to be classed among "miserable sinners", and rosy little rollicking lads and lasses, so full of frolic, so free of care.' Hartley, *Seets i' Blackpool, Fleetwood, Lytham, and Southport* (1890), 34.

> shoo wor happy. 'Just stop thear wol aw come back', aw sed, an' aw left her to explore what wor in a big buildin' at wor blazin' wi leet.[40]

These accounts challenge the idyllic description of the holiday resort as an inclusive space of happiness and the frequent carnivalesque theorization of the British seaside.[41] The failure of the resorts to match up to these ideals was clear to contemporaries, as we can see in the following description of Brighton:

> It is London repeated on a small scale, without the smoke and the slums, and with a purer atmosphere, though with scarcely less of a crowd. The shops are London shops, the actors at the theatre belong to London companies... *Like London, too, its social life is sustained by many classes circulating in a general current, but never affiliating*, and while in one quarter coronets are no rarity, and a prince leads society, in another the excursionist of a day, or the tradesman spending a two-weeks' holiday, smokes his brier or cutty and eats his shrimps without feeling the disparity, and without realizing that Brighton was not made especially for the delectation of his own class.[42]

The presence of class distinctions in contemporaneous accounts of holidaymaking has given scholars occasion to debate the applicability of concepts such as 'carnivalesque', 'utopia', or 'saturnalia' to Victorian and Edwardian holiday resorts and tourism in general.[43] Once we take into consideration the emotional meaning of the holiday this problem ceases to be merely theoretical. Contemporaries were not promised nor they were expecting the holiday resort to reverse social structures or to provide a temporal inversion of hierarchies; what was promised, though, was a special emotional experience. As a matter of fact, following one of Bakhtin's definitions of the carnival as 'the people's second life, organized on the basis of laughter' the concept could have been useful for understanding the place of emotions in leisure and holidaymaking. Unfortunately, scholars usually used

[40] 'Everything looked strange and grand to Mally, and as we sat down for a bit of a rest, I noticed that her face wore the same serene smile that I've seen on it when she's been siting listening to a sermon to one of her favorite texts, and I knew she was happy'. Hartley, *Seets i' Blackpool, Fleetwood, Lytham, and Southport* (1890), 34.

[41] See for instance Chris Rojek, *Capitalism and Leisure Theory* (Oxon: Routledge, 2013), 26–8; Webb, 'Bakhtin at the Seaside'; Urry, *The Tourist Gaze*, 29; Rob Shields, 'The "System of Pleasure": Liminality and the Carnivalesque at Brighton', *Theory, Culture & Society* 7, no. 1 (1990): 39–72; James Walvin, *Beside the Seaside: A Social History of the Popular Seaside Holiday* (London: Allen Lane, 1978), 63.

[42] Anon., 'The English at the Seaside' (1881), 488–9/PQHN. My emphasis.

[43] See for instance Robert Troschitz, 'Liminal Seaside? Working-Class Tourism in the Nineteenth Century', in John Storey (ed.), *The Making of English Popular Culture* (Oxon: Routledge, 2016), 104–17; Rob Shields, *Places on the Margin: Alternative Geographies of Modernity* (London: Routledge, 2013); Beetham, 'Lancashire Seaside Publications', 33; Webb, 'Bakhtin at the Seaside'; Borsay, *A History of Leisure*, 226–7; Cross and Walton, *The Playful Crowd*, 61–2; Shields, 'Liminality and the Carnivalesque at Brighton'.

the concept for the study of morality and class relations, and it was hardly applied in the field of the history of emotions.[44]

From the medical point of view this emotional experience was promoted as an antidote to the negative effects of 'work'. However, the depiction of the holiday in vacationers' accounts as well as in popular literature reveals that being 'off work' did not necessarily loosen the tensions related to social class. In fact, as Harold J. Perkin observes, 'the class-consciousness of the Victorians, which showed itself in the segregation by class and sub-class in the social zoning of towns and suburbs . . . was nowhere more evident than in their pleasure resorts'.[45] The medical discourse on the emotional experience of work, worry, and business tended to undermine the emotional experience of class. It is no surprise that middle-class professionals would not understand that the emotional burden of the lower classes was not so much about the 'work' but about class, and as their own accounts make apparent, this problem was more difficult to 'cure'.

Emotional self-reflection and the question of authenticity

All this is not to say that vacationers from all classes did not write about having fun and being happy on their holidays. Despite the gap between the professional understanding and definitions of the pathological effects of work and the experience of actual workers-vacationers, and although many of them, certainly members of the working classes, did not see doctors regularly, they were aware of the therapeutic purposes of holidaymaking, including its emotional aspects.[46] The quest for emotional and mental relief is present in popular seaside stories, poems, and reviews of resorts, and similarly to the professional literature, it is usually cast as relief from the effects of urban work. The 1880s and the following decades witnessed the development of several articulations of this narrative: when feeling 'worn with being busy', 'used up, or worked out', one should go to a watering place to 'recruit himself' and be 'repaid for toil and trouble'.[47] The painter Louise Jopling portrayed her ideal holiday by the sea, '[i]nto whose health-giving waters one may tumble unnoticed before the morning meal', by proclaiming '[w]hat a

[44] Mikhail Bakhtin, *Rabelais and His World*, trans. Helene Iswolsky (Bloomington: Indiana University Press, 1984), 8. The role of emotions in Bakhtin's analysis of the carnival is most apparent in his accounts of laughter and of 'the comic', e.g. 4–7, 305–6. See also Simon J. Williams, 'Modernity and the Emotions: Corporeal Reflections on the (Ir)Rational', *Sociology* 32, no. 4 (1998): 747–69. Emotions are mentioned in some 'carnivalesque' theorizations of the British holiday resort, but not developed as an analytical category, e.g. Cross and Walton, *The Playful Crowd*, 61–2; Webb, 'Bakhtin at the Seaside', 133.

[45] Perkin, 'The "Social Tone" of Victorian Seaside Resorts in the North-West', 180.

[46] Adams provides evidence also for the popular acquaintance with water cure methods treatments. Adams, *Healing with Water*, 92–3.

[47] Aliph Cheem [Walter Yeldham], *Lays of the Sea-Side* (1887), 6, 9; H. S. Forman, *A Holiday Tour in and around Whitby* (Bradford: P. Lund & Co., 1896), 3, 40.

relief to the fatigued brain and weary nerves' it was.[48] H. S. Forman, 'simply a business man out for a holiday', described how his 'pleasant memories' of 'one short day' in Whitby 'have lingered on through years of business troubles and anxieties', and advised his readers to go enjoy themselves at the same watering place in order 'to fight the coming struggles of life'.[49]

Even though one author claimed that '[i]t is a great mistake to suppose that enjoyment is expensive, or that happiness is costly', this seems to have indeed been a common assumption.[50] Aware of the emotional objectives and effects of the holiday and upset with its costs, some identified the emotional experience as the product essentially being consumed by holidaymakers. Hence, the consumption of 'health and jollity' for a price of 'sixpence or a shilling a time' was regarded by one author as a good 'investment', while another suggested that a worker who enjoyed a good holiday will be 'repaid for the time and money spent', so much so 'that he might almost be justified in entering it up as part of his business or professional expenses'.[51] A similar perception of investment is echoed in the words of a reviewer reminiscing about his past holidays, who explicitly linked the price of the holiday activities being consumed with the 'amount' and genuineness of the vacationer's emotional experience:

> These times have changed, we are too important now to content themselves [sic] with primitive Hornsea and its frugal enjoyments; we must go farther afield and spend more money. Yet, after all, I question if we extract more real happiness out of our wider and more expensive range of travels than we found in the simple enjoyments of 'ye olden time'—all the more because they were the best we could afford.[52]

Vacationers' accounts of their holidays often described the emotional experience not as a merely inner or individual moment, emphasizing instead its collective context. Writing about his visit to Blackpool, Howells noted: 'it is not the shipping that most or at all impresses the visitor at Blackpool; it is the myriads of other visitors . . . and whether you were a spectator of the throng or an infinitesimal particle of it the multitude was always astounding'.[53] The experience of the holiday crowd is just one example of the rise of consumerist leisure and the changes it generated in the late Victorian public sphere. An inherent element of mass leisure

[48] Anon., 'Some Well-Known Working Women's Ideal Holiday', *The Motherwell Times*, 23 July 1892.

[49] Forman, *A Holiday Tour In and Around Whitby*, 3, 35, 40.

[50] Anon., 'Nature', *The North-Eastern Daily Gazette*, 20 April 1889/BNA.

[51] Anon., 'Yarmouth in August', *The Comity of Middlesex Independent*, 29 August 1891, reprinted in: *Holiday Notes in East Anglia* (1894), 43; Musgrove, *Holidays and How to Use Them* (1914), 15.

[52] Anon., 'On My Back', *The Hull Packet and East Riding Times*, 25 June 1880/BNA; see also Benger Alaluf and Illouz, 'Emotions in Consumer Studies'.

[53] Howells, 'The Waters of Blackpool' (1911), 874.

activities and spaces (e.g. the music hall, sport events, theatre) was the gaze of the public and the impression that the holidaymaking crowd 'belongs as much to the "sights" of the town as do the sands, the promenade, and the palaces of pleasure'.[54] Frequently, these accounts included emotional descriptions of the crowds, such as 'ivverything lukt lovely an' ivverybody happy'; '[t]he merry visitors seem lithe and gay, [a]nd much enjoy the pleasant promenade;' 'all were young and looked glad'; 'I have seen... the infinite satisfaction of the gaping multitudes on shore', to mention a few examples.[55]

The crowd played a crucial role in the emotional experience of the holiday as well as in the process through which vacationers learned to express their emotional engagement. In many cases, authors described not only the emotional atmosphere, but evaluated, and sometimes judged, the causes and the authenticity of the emotional engagement of fellow visitors. Observing the middle-class visitors on the North Pier, Ab-o'-th'-Yate mentioned that '[t]hey con raise a lot o' fun eaut o' nowt, an' laaf at things ut others wouldno' open their een at', and later noted again that they were 'laafin' at nowt'.[56] Even Sammywell Grimes, who appears quite satisfied throughout his holiday trip, noted that the people he met strolling on the promenade at Blackpool 'wor all smilin' an' enjoyin' thersen', an observation that made him wonder whether they were not actually wishing that 'they wor at hooam mindin' ther business asteead o' wastin' ther time as they wor dooin'.[57] Similarly, in John Oxenham's *The Very Short Memory of Mr. Joseph Scorer and Other Seaside Experiences*, one of the protagonists admits that he 'was doubtful at times' whether the children were enjoying 'as much as they most certainly seemed to'. He himself, though, 'had a genuine day of rest, and enjoyed it with a novel sense of freedom from the cares and worries of life'.[58]

The question of emotional genuineness was raised specifically by American writers. '[T]he English sea-side', wrote an anonymous reviewer, 'is crowded with apparently satisfied pleasure-seekers.' The same writer also criticized '[t]he vulgar

[54] Anon., *A New Pictorial and Descriptive Guide to Blackpool, Lytham, St. Anne's, Fleetwood, and The Fylde* (1902), 16. See also Borsay, 'Visualising the Seaside, c. 1750–1914', 181; Charles Baudelaire, *Paris Spleen*, trans. Louise Varese (New York: New Directions Books, 1970 [1869]), 20; Benjamin, *The Arcades Project*, 21; Susan Buck-Morss, *The Dialectics of Seeing: Walter Benjamin and the Arcades Project* (Cambridge: MIT Press, 1991), 81.

[55] 'Everything looked lovely and everybody happy'. Hartley, *Seets i' Blackpool, Fleetwood, Lytham, and Southport* (1890), 27; Anon., *Confessions of a Coquette, while Staying at Scarboro', Whitby, & Bridlington* (1888), 13; Howells, 'The Waters of Blackpool' (1911), 878; Anon., 'On My Back', *The Hull Packet and East Riding Times*, 25 June 1880/BNA.

[56] 'They can raise a lot of fun about nothing, and laugh at things that others wouldn't open their chin at; like snails from one end to the other, and laughing at nothing'. Ab-o'-th'-Yate [Benjamin Brierley], *Drop't on at Blackpool* (1881), 14–5.

[57] 'Were all smiling and enjoying themselves; they were at home minding their business instead of wasting their time as they were doing'. Hartley, *Seets i' Blackpool, Fleetwood, Lytham, and Southport* (1890), 93–4.

[58] John Oxenham, *The Very Short Memory of Mr. Joseph Scorer and Other Seaside Experiences* (London: Simpkin, Marshall, & Co., 1903), 34.

excursionist—the London 'Arry', stating that 'his mirth is not genuine'.[59] Likewise, referring to the British 'merry-makers', Howells wondered whether 'it was merry they were making'.[60] Generally, Howells had very high expectations when gazing at the crowds:

> [T]he English do not know how to bathe, nor how to be picturesque. They go to the sea-side in their overcoats, and wear dark colors, which are not pleasing to the spectator, and which give the throngs a penitential seriousness of aspect. The spectator has a claim to be considered in the matter: good society is always expected to be interesting... But the sparkle, the animation, the light-heartedness, and the variety of costume which are seen in America, and which contribute to a dramatic or, if you will, theatrical effect, are missing at the English sea-side, and an aspiring play-writer could not gather the faintest inspiration from it in his most impressionable mood.[61]

This text exposes holidaymakers' awareness of the theatrical function of other consumers and the conscious role of themselves as spectators.[62] Some examples also reveal that holidaymakers were aware that others were judging the authenticity of their own emotional states and expressions, like one writer who realized that his companion was 'hinting, with an obvious sense of injury, that I'm not cheerful, buoyant, or gay'.[63] When on holiday, so it seems, one had to comport oneself in a way that fit the image of the happy holidaymaker. This emotionally performative dimension of holidaymaking was enhanced by the disciplining power of the crowd. As Angelia Poon argues, the negotiation 'between positions of being holder of the gaze and object of the gaze' was central to the process of the embodiment and performance of 'Englishness'.[64]

Writing about Victorian sensation theatre, Lynn Voskuil shows that starting in the 1860s, audiences both consumed sensations at the theatre and at the same time learned how to express their emotional engagement with the play within a public space. Interestingly, this experience of enacting codified emotions and sensations was nonetheless perceived as individual, spontaneous, natural, and authentic.[65] Voskuil's study shows that spectators 'knew how and what they were supposed to feel without relinquishing the sense of shock and spontaneity that was part of the experience of viewing sensation drama'. Holidaymaking underwent a similar process when transforming into a commercialized emotional experience. Unlike

[59] Anon., 'The English at the Seaside' (1881), 484, 486/PQHN.
[60] Howells, 'The Waters of Blackpool' (1911), 878.
[61] Anon., 'The English at the Seaside' (1881), 484, 486/PQHN.
[62] Edensor, 'Performing Tourism, Staging Tourism'.
[63] Anon., 'Bored at the Sea-Side', *The Observer*, 20 August 1871, 5/PQHN.
[64] Angelia Poon, *Enacting Englishness in the Victorian Period: Colonialism and the Politics of Performance* (Farnham: Ashgate Publishing, Ltd., 2008).
[65] Voskuil, 'Sensation Theater, Commodity Culture, and the Victorian Public Sphere', 261–2.

the playgoer, though, the holidaymaker was simultaneously the 'spectator' and 'the show'. In the absence of the 'fourth wall', the emotional expressions of holidaymakers were more than 'signs of their engagement, of their participation "as if" they are present in the scene';[66] they were a manifestation of performative consumption itself.

The perception of emotion as a universal category facilitated holidaymakers to apply a pseudo-scientific gaze over fellow visitors and analyse the origins of the emotional experience.[67] This objective was articulated in a surprisingly straightforward way by 'Lucy':

> 'For the majority of the people who attend these Fairs, sound provides their rational amusement. "Make a noise, you chaps," cries one; and a chorus of song, some coarse witticisms, a hoarse laugh, a slap on the back, the lifting of feet in a rude dance, follows as a matter of course. There is nothing tuneful in the song, nothing melodious. But these movements, these utterances, contain pleasure for the people who give rise to them; and an examination and analysis of this enjoyment occasions the belief that for them the pleasures of a recreation exist in ratio to the scope the recreation possesses for the production of noise. Hence the volume of sound which rises and falls in intensity, but is always dominant.'[68]

Josephine Kane brings a similar example of a journalist who in 1908 analysed the emotional experience at the amusement park, arguing, for example, that the roller-coaster ride allows 'the modern man' to enjoy 'primitive emotions in a scientific fashion'.[69] Voskuil argues that the new curiosity of middle- and upper-class people about allegedly natural and authentic emotional experiences evoked during leisurely activities made them more tolerant towards the otherwise condemned habits of the lower classes. In effect, this meant that the shift towards consuming emotions as part of leisure activities had the potential to challenge social boundaries.[70] Voskuil's analysis can also be applied to the influence of the holiday on class relations as it is reflected in descriptions of the emotions of holiday crowds toward the turn of the century and onward. 'Of course we frankly admit that occasionally there is an invasion, just as there is at every other seaside resort of holiday-folk who are determined to compress as much enjoyment as possible into the few hours at their command', stated an article about Southend-on-Sea. But, the authors continued, 'if some of these holiday makers—honest and

[66] Ibid. 261–2.

[67] Cf. Tim Edensor's distinction between unreflexive and reflexive dispositions and their effect on tourist performance, Edensor, 'Performing tourism, staging tourism', 62–3.

[68] Anon., Ladies' Column by 'Lucy', *The Whitby Gazette*, 2 August 1907, 7/BNA.

[69] 'A Fortune in a Thrill', *Sunday Chronicle* (Manchester), 23 August 1908 quoted in Kane, 'Edwardian amusement parks: the pleasure garden reborn?', 232.

[70] Ibid. 251–2, 262.

"hard-working" East-enders most probably—are not exactly gifted with the charm of manner and repose of Lady Clara Vere de Vere, they nevertheless enjoy themselves thoroughly in their own way, and do not in the least interfere with the tastes and pleasures of visitors of another stamp.'[71] This text articulates an ethical argument that also betrays a specific understanding of the emerging prominence of emotional well-being. The authors not only recognized that different classes enjoy themselves in different ways, but acknowledged that all classes had the right to pursue pleasure and joy in their own 'style' without the intervention of other classes.

Another author, writing about Yarmouth, noted that the behaviour of some visitors, 'though not pleasant to better educated folk, need not interfere with their pleasure', and asserted that '[t]he space is large and "birds of a feather flock together"'.[72] A similar demand to respect and even marvel at the habits of the working classes appeared in an 1894 article about Blackpool, in which the author, referring to the 'exceedingly plebeian character of the visitors', explained that

> [t]o the average person of an open mind this very feature is the one that most recommends the town to notice. What can be more cheerful to the humanitarian than the consciousness that these young sparks in straw hats and deerstalkers, with yellow shoes to their feet, and jewelled rings to their hands—who crowd the promenade from morn to midnight—are hard-working artisans when they are at home? It is the same with the feathered and flounced damsels, who laugh loudly as they take the ozone to their lungs. They are factory girls for about three hundred days in the year. Modern progress and the rail way companies enable them all, periodically, thus to wrap themselves round with the sweet illusions of temporal greatness... Blackpool is the most vulgar seaside resort in the country, and therefore one of the most important, prosperous, and remarkable.[73]

The call for accepting the behaviour of working people on the grounds that they were simply looking after their own—justified—emotional needs also appeared in a book written by a medical expert. 'There are many thousands of toilers who find other places no more crowded than the humble homes in which they live, and to people of this type such discomforts as we have enumerated do not detract from the pleasures of the day', he explained, adding that '[i]t may be that they even add to them on the principle, "The more the merrier"'. The doctor then asks the reader to 'let them go and enjoy themselves, but let other and more fastidious people stay

[71] Anon., 'Southend-on-Sea', *Civil Service Guardian*, June 1890, reprinted in: *Holiday Notes in East Anglia* (1894), 53.

[72] Anon., 'Yarmouth in August', *The Comity of Middlesex Independent*, 29 August 1891, reprinted in: *Holiday Notes in East Anglia* (1894), 43–4.

[73] Anon., 'The Sweets of Popularity', *All the Year Round* 11, no. 271 (1894): 225–8, 226.

at home'.[74] William Dean Howells' account serves again as a nice example. Although the review as a whole is very critical and negative, the encounter with the British working classes induced in the author curiosity and even empathy. Interestingly, his curiosity relates mainly to the question of enjoyment, given that the crowds seem to be having a good time, while he was certainly not:

> Of course we wondered, or at least we said we wondered, why all those poor souls should have come to Blackpool for what we could see of their pleasure, but a little reflection ought to have taught us to moderate our amazement. People even so wrong-headed as we like to think the English average do not take their year's savings in one hand and their families in the other and hurry summer after summer for one hot week on or behind those sands out of mere midsummer madness.[75]

Puzzled with the enjoyment of these people, whom he at first simply could not understand, Howells then raised a hypothesis that echoes the writings of Bourdieu sixty years later:

> Perhaps if we had seen the towns and the houses which they lived in the rest of the year we might have found their joy (it did not look much like joy, though) in their Blackpool sojourn more justifiable than it appeared. I am afraid we did not consider, either, the gratification of their natural pride in being in Blackpool, and of meeting people they knew there, whom afterwards they could exchange reminiscences with, and remind of the splendid times they had there together... Poor, ordinary fellow men and women, how few such moments could come to them even at Blackpool, the sole romance of their workaday year, on which they did well to lavish the savings from their hard earnings! In spite of their wastefulness it must have been much to them that they could have so much of that joy for nothing.[76]

However, as if to reassure us that the social borders remained very clear even at the seaside, Howells included in his text a sad confession:

> Even under cover of the dark I had not the courage to ask any of them if it were not so. I knew very well they would be glad to say, or to talk on any terms, those fathers and mothers of families, if only to contradict one another in the report of their common opinions and emotions. But travel is mostly a waste of opportunity. You meet innumerable people who would eagerly tell you their lives or analyze their characters to you, if you would offer them the chance by the

74 Musgrove, *Holidays and How to Use Them* (1914), 197.

75 Howells, 'The Waters of Blackpool' (1911), 877.

76 Ibid. 877–9.

> slightest question, and yet you pass them dumbly by while inwardly you are hungering and thirsting to know about them. How rich I might have made this page if, when we venturously formed part of the organism on the quay, I could have asked some elbow neighbor what he was when he was at home, and how long he was going to stay, and how he liked it; and was that lady with him his wife; and how many were they in family. He might not have responded in the same ingenuous spirit, but I think he would.[77]

Emotion knowledge, guidance, and responsibilization

The performative character of the holiday could not be sustained by the disciplining gaze of the crowd alone.[78] It also presupposed the existence of knowledge about emotions and how they ought to be managed and expressed. Previous chapters demonstrated that the medical literature was known among the political and economic circles involved with the holiday industry, but just as important is the fact that this knowledge was available to holidaymakers, too. I have mentioned a number of examples in which the therapeutic narrative appeared in holiday popular literature, sometimes in a straightforward manner, sometimes with a hint of sarcasm. However, this knowledge was also articulated in an instructional manner, in both professional and popular literature expressly written for the holidaymaker.

Medical advice literature

'There is an art of holiday making which needs to be acquired', declared the author of an 1871 article titled 'Holidays', without which the holiday might 'turn out neither very pleasant nor very profitable'. The author stressed that especially 'since we are all so profoundly convinced that holidays are a necessity of nineteenth-century life, and mean to take them at any rate, it is worthwhile to spend a little trouble learning it'.[79] Holidaymaking carried a risk of failure, and for those who advocated it for therapeutic goals, this meant more than just disappointment: a bad holiday could be dangerous. It is for this reason, the writer explained, that 'study and consideration' are required in order to guarantee that the holiday would be truly beneficial.[80] In other words, holidaymaking needed to be learned. After all, not everyone 'can feel all this innocent, healing, strengthening pleasure without affectation', as some people 'are so completely absorbed in town life and so dulled and deadened by workshop influence that their finer sensibilities die

[77] Ibid. 879. [78] Edensor, 'Performing tourism, staging tourism', 72.
[79] Anon., 'Holidays', *The Bolton Chronicle*, 16 September 1871, 5/BNA. [80] Ibid.

out'.[81] Moreover, those who 'leave their homes in search of change of air, either for health sake or relaxation, have generally anxieties enough, inherent to the change', and thus should avoid 'having them multiplied by fears lest the resort which they have selected may bring upon them evils that they know not of'.[82]

The risk of failure generated demand for guidance. Whereas in 1888 people had to 'actually consult a physician'[83] before choosing a holiday destination, by 1914 they could turn to the growing body of literature designed to support the holidaymaker.[84] Thus reported the *Daily News & Leader*:

> The day of the careless, happy-go-lucky holiday is past. We have all become exceedingly self-conscious about our recreations and vacations. We doubt and diagnose and discuss from medical, physiological, and psychological standpoints before deciding whether Brighton or Bournemouth shall be our goal. And of the making of instructive articles and books telling people what to do when on holiday and what not to do, there is no end. The latest is 'Holidays and How to Use Them', by Charles Musgrove, MD, published at half-a-crown by J.W. Arrowsmith—a most comprehensive survey of the whole question.[85]

The newspaper then directly quoted parts of Musgrove's book. Previous chapters discussed Musgrove's book *Nervous Breakdowns and How to Avoid Them* (1913), which included a short chapter on the subject of holidays that noted that 'it is well worth while to consider how the time and money involved may be expended to the best advantage'.[86] A year later Musgrove dedicated a whole book to the subject. 'There is no reason', he argued, 'why the health-seeker should not get plenty of enjoyment out of his vacation, or why those who set out with the sole intention of amusing themselves should not benefit in health at the same time.' Thus, the aim of the book was to help the holidaymaker in 'combining a maximum amount both of pleasure and health'.[87] Musgrove's book is a remarkable example of the secondary economy of popular holiday advice literature established by medical professionals, a genre that should be seen as a predecessor to the popular psychological advice literature that flourished in the 1920s, in which ostensibly scientific knowledge was interwoven with moral guidance.[88]

[81] Anon., 'Nature', *The North-Eastern Daily Gazette*, 20 April 1889/BNA.

[82] Haviland, *Scarborough as a Health Resort* (1883), 80.

[83] Anon., 'Health and Holidays (extract from The Christian Million)', *The Star Guernsey*, 8 September 1888/BNA.

[84] On hydrotherapy and naturopathy advice manuals and their 'agenda of self-help', self-care and self-management see Adams, *Healing with Water*, 13, 91–9, 167.

[85] Anon., 'Holidays for the Nervy', *Daily News & Leader*, 10 April 1914/NYCRO DC/HRG/BPC 1914.

[86] Musgrove, *Nervous Breakdowns and How to Avoid Them* (1913), 138.

[87] Musgrove, *Holidays and How to Use Them* (1914), 10, 11.

[88] Eva Illouz, *Saving the Modern Soul: Therapy, Emotions, and the Culture of Self-Help* (Berkeley: University of California Press, 2008), 52.

Under the guidance of medical authority, holidays were regulated and rationalized just like working time: 'Get up early by all means'; 'Reduce your newspaper reading'; 'Business communications are of course taboo, and the fewer letters you get the better it will be for you'; 'Do not adopt the superior attitude of the man from town'; and above all: 'Do not be disappointed.'[89] The book even included a section on 'How to do nothing', in which the vacationer was warned: 'Do not suppose, however, that you are going to get much pleasure out of it by simply mooning around and gazing about you in a desultory, aimless manner.'[90] Indeed, it is impossible to definitively know whether these instructions were followed or how they influenced the feelings of their readers. However, such texts do shed light on the emotion knowledge that was available to contemporaries and that contributed to the formulation of emotion norms.[91] Additionally, it reveals at least one way in which the emerging body of knowledge about 'change' and emotion management was made accessible to potential holidaymakers. *Holidays and How to Use Them* informed the reader about the importance of one's emotions to health and provided practical guidance on how to manage them through holidaymaking. Musgrove explained to his readers that when a man 'goes on from day to day, feeling that he has not the energy to devise new plans or strike out on new lines', he should realize that 'he has reached this stage that a holiday will refresh him as nothing else can do'.[92] In his opinion, individual feelings function as an inner compass that lead one to the source of health: 'The wind comes off the sea laden with life-giving properties. It is no imagination. You do not need to look for the direction of the breeze, for your own feelings will tell you whether it is blowing off the land or off the sea.'[93] Furthermore, emotions serve as a litmus test for detecting the state of one's health:

> Sunburn and health are inevitably associated together in our minds, simply because we have found by experience that we are all different beings after a sunny holiday. We not only feel stronger and hardier, but worry less, fidget less, and take things more as they come. A cheerful philosophy and good health also go together. The light-heartedness of the Latin races is largely a question of the climate in which they live.[94]

The emotional self-reflection demanded by such literature was geared towards promoting practices of self-management and the optimization of personal well-

[89] Musgrove, *Holidays and How to Use Them* (1914), 29–30, 31, 60, 126. [90] Ibid. 51.

[91] Pascal Eitler, Stephanie Olsen and Uffa Jensen, 'Introduction', in Ute Frevert et al. (eds), *Learning How to Feel: Children's Literature and Emotional Socialization, 1870–1970* (Oxford: Oxford University Press, 2014), 7–8; Pernau, 'Emotions in Indo-Muslim Advice Literature'; Illouz, *Saving the Modern Soul*, 52.

[92] Musgrove, *Holidays and How to Use Them* (1914), 15. [93] Ibid. 136–7. [94] Ibid. 31.

being. The text also defines which practices certain readers can use to achieve a particular emotional state: an urban holidaymaker 'may lift his eyes to the jewelled heavens and feel his mind transported beyond the cares and worries of his daily life'; holidaymakers from the countryside are advised 'to walk along streets where they know no one, and where no one knows them', because '[i]t gives them a sense of freedom, such as that which a townsman feels when he looks out over an expanse of ocean or some wide-spreading landscape'.[95] In contrast to his strict differentiation between the urbanite and the villager, Musgrove's approach was quite egalitarian when it came to gender. He claimed that some feelings were in the 'nature of every man and woman', and even stressed that '[m]en would do well always to remember that a woman needs a rest from the worries of housekeeping quite as much as they themselves need a respite from their work'.[96]

The potential experience of boredom that was discussed in the previous chapters was also one particular emotional state targeted by Musgrove. Whereas the resort industry countered the risk of boredom by providing more entertainment, the advice literature took a more complex approach. Musgrove argued that 'the tired, nervy people, who wish to have their minds recast and leave their cares behind them', should not look for a quiet and restful holiday resort, for in such places 'they are apt to become bored and their last state is worse than their first'. The author explained that although the 'common idea in planning a restful holiday is to seek some quiet spot...and do nothing', in reality '[t]his is just what a tired brain cannot do'. According to Musgrove, such a person 'has got into a state of perpetual motion and will not stop worrying'. It is therefore that when choosing a holiday destination, tired and worried people should find 'a place where they can take things easily, and yet have enough going on before their eyes to keep their minds occupied'. *Holidays and How to Use Them* even contains a section titled 'Remedies for boredom'.[97]

Musgrove's book is, to the best of my knowledge, the first full-length book exclusively dedicated to holiday advice. However, holiday advice from doctors addressing holidaymakers, regardless of whether they were 'ill', appeared scattered in the general press much earlier. Such contributions did considerable work in positioning the medical profession as the expert in achieving 'healthy pleasure'.[98] One intriguing characteristic of both early and late holiday advice texts is their pairing of concrete injunctions with very ambiguous conclusions. In practice, most authors tended to leave the main decision in the hands of the readers who, it seems, were assumed to be capable of a very high level of self-reflection.

[95] Ibid. 146–7. [96] Ibid. 12, 48, 146–7. [97] Ibid. 24–5, 45, 46.

[98] See, for instance Anon., 'Holidays', *The Bolton Chronicle*, 16 September 1871, 5/BNA; Andrew Wilson, 'Holidays: For Pleasure, Rest and Health', *Cheltenham Chronicle and Graphic*, 28 June 1902, 7/BNA.

Worrying and busyness, for instance, appear as major reasons both for and against taking a vacation:

> At times of course it does happen that other reasons than those connected with money may make a man's mind uneasy when he is thinking about taking his holiday. Business is in a particularly critical condition, or your confidential assistant is ill, or something or other makes you feel that you ought to be at your post, and that things are likely to go wrong if you are not. In any such case it is better to defer one's holiday than to take it, for even if the dreadful contingency does not come, the thought of it will disturb you quite sufficiently to prevent your getting any good . . . A single week of absolute oblivion to all consideration bearing upon business will do more to recruit one's energies than a couple of months of hybrid holiday-making, neither work nor play.[99]

More generally, 'those who cannot stand much fatigue or discomfort' will do better staying at home, as '[t]here is enough trouble in the world already without adding to it the manifold inconveniences of travelling at such times, with crowded trains, crowded restaurants, crowded everything'.[100] The individual holidaymaker was responsible for undertaking the emotional diagnosis that would help him or her determine whether she should take a holiday or not and for choosing the destination that best suited her emotional state. The literature on how to make such decisions usually argued with the language of pathology.[101] Considering whether to choose a visit to the mountains or to the seaside, MD Isaac Burney Yeo argued: 'Highly irritable, nervous organisations, people who, as we say, take too much out of themselves, profit more by mountain than by sea air. For those, on the contrary, who have no tendency to nervous irritability, and who are in a condition to bear the increased stimulus to tissue-change, sea air is a more powerful restorative agent.'[102] Some resorts were considered particularly beneficial for people suffering 'from mental and physical debility, the consequence of prolonged application to business or study', others for those 'who suffer from chronic dyspepsia or nervous relaxation', or from 'various forms of nervous exhaustion so common to those whose lives are subject to a constant mental strain'.[103] Planning the optimal holiday thus required a process of self-reflection and self-pathologization.

[99] Anon., 'Holidays', *The Bolton Chronicle*, 16 September 1871, 5/BNA.

[100] Musgrove, *Holidays and How to Use Them* (1914), 197.

[101] Vladimir Jankovic observes a similar trend earlier in the century, see Jankovic, 'The Last Resort: A British Perspective on the Medical South'.

[102] Yeo, *Climate and Health Resorts* (1890), 29.

[103] Anon., *My Holiday: Where Shall I Spend It?* (1878), 91; Anon., *The Dictionary of Watering Places, Seaside and Inland, at Home and Abroad* (London: L. Upcott Gill, 1881), 37; Cole (ed.), *Hastings and St. Leonards-on-Sea as a Health and Pleasure Resort* (1884), 37.

In this way, in addition to formulating ideas on emotions and how to manage them, the holiday advice literature helped shape the emotional consumer as a person able to determine her or his own emotional needs and the appropriate means for achieving them. At the same time, by supplying people with expert knowledge, this literature added one more voice to the responsibilizing *ethos* of the emerging self-help culture, framing failure and success as being dependent on individual choices.[104] That failure and success were seen as a matter of individual choice was expressed very clearly in the holiday literature of the early twentieth century. One author admitted that sometimes holidays offer no benefit, and that some people require 'a week's rest to recover from the effects of a day's holiday'. However, he maintained that no general conclusions should be drawn from these individual cases, and argued that the advantages of the bank holiday must not be put into question only '[b]ecause a man or a woman here or there abuse their opportunity and their self-respect'.[105] A similar argument was raised by MD Musgrove. He too noted that some people do not enjoy their holiday 'and return home more tired than when they set out', but insisted that: 'the great majority of persons are able bodied, and when they complain of lassitude and other symptoms during a seaside holiday it is usually because of conditions for which they themselves are responsible'. Holidaymakers, Musgrove believed, should know how to base the choice of their holiday destination on the reasons why they needed a holiday in the first place. 'If people who are tired out with the noise and racket of a large town choose some popular resort', that is, a resort with 'endless promenades, crowds of people, bands and motor char-à-bancs', then 'they must not be surprised when they find their nervous symptoms worse than they were at home'.[106]

Emotion knowledge by holidaymakers

The emotion knowledge available to holidaymakers was not only delivered in medical advice texts, but also through popular literature, like stories and poems about watering places and personal accounts written by vacationers. These texts made sense of the emotional experiences that could be had away from home and offered advice on how to stimulate and express them.[107] In her study of Lancashire seaside publications, Margaret Beetham argues that seaside stories went beyond

[104] Illouz, *Saving the Modern Soul*, 152–3. For an account of similar, yet accelerated processes in neoliberal societies, see Cabanas, 'Rekindling Individualism, Consuming Emotions'.

[105] Anon., 'Bank Holidays', *The Western Times*, 5 April 1904, 7/BNA.

[106] Musgrove, *Holidays and How to Use Them* (1914), 137–8.

[107] Similarly, Edensor notes that 'guidebooks are a kind of master script for tourists which reduces disorientation and guides action', I would agree and add that such texts also guide their readers *what and how to feel*. Edensor, 'Performing tourism, staging tourism', 73.

their more obvious function of representing holiday experiences by contributing to the definition of holidaymaking and the significance attributed to it.[108] While Beetham focuses on the meaning-construction of time and space, these popular publications also informed the meaning of emotions and their relation to holidaymaking.

Holiday stories and poems usually narrated the experience of the protagonist, relating his or her activities along with a description of their emotional impact. In this sense, and in contrast to the holiday advice texts, which cannot tell us much about the actual practices of their readers, popular literature can be read more as a kind of ego-document. Holiday stories were strikingly similar to resort publicity in both content and form, as depicted in Figures 5.1–5.4, which show popular stories and guides side-by-side. The plot, be it about a romantic encounter or a series of comical incidents, was usually structured in a way that introduced the different attractions of the resort, even going so far as to include prices of entry. In this way, popular stories functioned as visitor guides that combined the plain listing of potential activities with their experiential dimension.

Under the title 'The Tobogganing and Switchback Railway', the author of *Adventures and Sketches at New Brighton* took up the role of travel guide, providing the reader with information about the attraction, its price, and the emotional experience attained:

> There's not a livelier feeling, than o'er each sense comes stealing,
> When in the great Toboggan slide, just erected on the shore,
> To all, my best advice is, for nothing half so nice is,
> And the very modest price is only twopence, nothing more;
> 'Tis a pastime that the doctors recommend to many a score,
> So toboggan evermore.[109]

Mentioning the doctors' recommendation, although probably sarcastic, nonetheless affirms that the therapeutic conception of holidaymaking was widespread. Worthy of the bard's attention, too, were the emotional objectives and significance of the holiday, as emphasized towards the end of the story:

> I've not describ'd one half the things, at
> every turn sure something brings,
> Congenial pleasures to the happy thousands
> on this Cheshire shore.[110]

[108] Beetham, 'Lancashire Seaside Publications'.

[109] Short, *Adventures and Sketches at New Brighton* (1888), 10.

[110] Ibid. 15.

A similar combination of a recommendation based on personal experience and practical information can be found in Laycock's *Fun on the Pleasure Beach*. The narrator of the opening poem, which lists off in succession the resort's numerous entertainment facilities, provided the reader with personal advice and recommendations (e.g. 'Yo' munno skip this'; 'I'd advise yo' to explore it').[111] The author then encouraged readers to share their own experiences and recommendations:

> Yo' may roam all o'er England,
> O'er th' wide world may seech,
> But yo'll find nowt that's grander
> Nor th' famed Pleasure Beach.
> Tawk abeawt fun an' frolic,
> An seets of aw sooarts!
> ... Th' Wayter Chute gives sensations
> Yo' conno weel miss;
> If yu' want a rare treat, folks,
> There's th' new Velvet Coaster,
> 'At glides up an' deawn.
> Swiftly, smoothly, an' sweetly;
> Ther's nowt like it in th' teawn.
> Th' River Caves an' their wonders
> One conno describe;
> ... I'd advise yo' to explore it—
> Keep th' seets to yoresel,
> Then tell others to go o'er it—
> I've been theer mysel'![112]

After the poem, the story begins, with Sally describing her holiday adventures with Sam, Jimmy, and Jane Ann:

> We aw went on a trip to Blackpoo', us lasses donned up loike duchesses, an' we spent every bit o' th' day at th' Pleasure Beach, Seawth Shoore. Eh! What gam we had. Yo' never seed th' loike i' aw yo're days. But I'll tell yo' abeawt it.[113]

[111] 'You must not skip this'. Laycock, *Fun on the Pleasure Beach* (1909), 8.

[112] 'You may roam all over England / Over the wide world may seek / But you'll find nothing that's grander / Than the famed Pleasure Beach / Talk about fun and frolic / And sights of all sorts! ... The Water Chute gives sensations / you cannot well miss / If you want a rare treat, folks / There's the new Velvet Coaster / That glides up and down / Swiftly, smoothly, and sweetly / There's nothing like it in the town / The River Caves and their wonders / One cannot describe ... I'd advise you to explore it—Keep the sights to yourself, Then tell others to go over it—I've been there myself!'. Ibid.

[113] 'We all went on a trip to Blackpool, us lasses dressed up like duchesses, and we spent every bit of the day at the Pleasure Beach, Sea Shore. Eh! What a fun time we had. You never saw the like in all your days. But I'll tell you about it'. Ibid.

Penny Guide to Scarborough. 27

RIDES AND DRIVES.

For detailed list of these see next page.

Two Shillings an Hour.

Riding, or attempting to ride, is a great diversion of excursionists, who rarely get outside of the Borough boundary with their steeds, and sometimes finish with an unintentional but economical sea dip.

Those, however, who can ride or drive abroad will find it easy to make in every direction both healthful and interesting excursions, whether they take a Southern course towards Filey and Bridlington, with the sea to

No Extra Charge.

the left of them, or a Northern course towards Whitby, with beautiful peeps at the sea on the right, or plunging inland, seek out Forge Valley, by way of Falsgrave and Seamer, or by way of Lady Edith's Drive, or Newby, Scalby, and Everley, they cannot go wrong. Waggonettes run

Figure 5.1 'Rides and Drives', from *Adey's Annual Popular Guide to Scarborough* (1887)

10 *Adventures and Sketches at New Brighton.*

THE TOBOGGANING AND SWITCHBACK RAILWAY.

THERE'S not a livelier feeling, than o'er each sense comes stealing,
When in the great Toboggan slide, just erected on the shore,
To all, my best advice is, for nothing half so nice is,
And the very modest price is only twopence, nothing more;
'Tis a pastime that the doctors recommend to many a score,
So toboggan evermore.

THEN a change of locomotion, is the Switchback Railway notion,
At the Liver' exhibition it made quite a great furore,
Here too it meets with much success, for speed of course they don't profess
To beat the Wallasey express, bringing its thousands to the shore,
Still whilst it lasts the fun is fast and furious too, I'm sure,
To please you evermore.

Figure 5.2 'The Tobogganing and Switchback Railway', from Short's seaside story *Adventures and Sketches at New Brighton* (1888)

don't slat thy sarcasm at me. Who's ever yerd of a' angel coed 'Jane Ann'?"

"Or who ever yerd o' one coed Sally?" Jane Ann snapped.

"That's enoof," said sensible Sam, "Drop it; we'll aw fly an' be cherubins for a change; it's cheap enoof."

WATER CHUTE.

So we climbed up th' stairs an' were put into another kind o' little booat wi' pointed nooases an' windmills whirlin' at oather end.

"Howd tight!" said th' gaffer, so this time Sam act'ally hed th' sense to grab howd o' me.

5

Figure 5.3 'Water Chute', from Laycock's seaside story *Fun on the Pleasure Beach* (1909)

In the following, Sally details their experience in each one of the amusements: 'we went on th' Wayter Chute'; 'We'll go on th' Seenic Railway neaw'; 'We went on th' Skatin' Rink'; and so on.[114] The story at once declares amusement and fun to be

[114] Ibid. 3, 6, 12.

From Photo by] THE GIGANTIC WHEEL. [*A. Dearden, Blackpool.*

Figure 5.4 'The Gigantic Wheel', from Blackpool Corporation's guide *Blackpool the Unrivalled Sea Side Resort for Health & Pleasure* (1898)

the objective of the holiday, feeds the imagination and expectations of the readers, and provides them with information about how the activities feel and how these emotions and sensations should be expressed: when to laugh, when to scream out of excitement, when to be sad (when the ride is over), how to articulate the bodily sensation of thrill ('My heart were i' my meawth').[115]

Samuel Laycock, one of the most popular poets of the Lancashire working class, also wrote about the Blackpool seaside holiday. The following extract from 'The Blackpool Fete'[116] features another iteration of the therapeutic perception of the holiday. Laycock addresses the 'poor childer o' toil' and calls on them to leave the polluted industrial city and go to the seaside to consume pure air:

> Come an' look at th' owd ocean, poor childer o' toil,
> An' rest yore tired limbs at th' seaside for awhile.
> When here yo'll get strength for engagin' i' th' strife
> 'At awaits every one i' th' great battle o' life.
> Leave thoose dark, smooky teawns, an' come drink o' th'
>   pure breeze
> 'At comes sweepin' so freely across the great seas.[117]

The last stanzas of the poem reinforce the idea that pleasure and health are inseparable. These stanzas also adopt the philosophy of responsibilization prominent in the medical advice literature, arguing that one's happiness depends on one's own practices, which, in this context, means going to the coast. Additionally, Laycock too offers practical guidance, such as on how to reserve accommodations, and mentions the economic burden of the quest for health and happiness by reminding the reader to bring a lot of money:

> Leave yore business an' books, don yo' up i' yore best,
> An' come au' watch th' sun as he sinks into th' west;
> ... Is it wise on yo' t' sacrifice pleasure an' health
> For that nick-named, perishin' bauble called wealth?
> Nowt o' th' sort; for yo'll never be happy, yo'll find,
> Unless yo'v good health an' a contented mind.
> Well, we think we've said plenty for one time at least—
> We've invited yo' o'er to a very rich feast;

[115] Ibid. 4, 6, 8. 'My heart was in my mouth'.

[116] Samuel Laycock, 'The Blackpool Fete' (Blackpool: The Times, undated)/BLP/Laycock, Samuel. Poems 1871–1908/LT51 Pamphlet.

[117] 'Come and look at the old ocean, poor children of toil / And rest your tired limbs at the seaside for a while / When here you'll get strength for engaging in the strife / That awaits everyone in the great battle of life / Leave those dark, smoky towns, and come drink of the pure breeze / That comes sweeping so freely across the great seas'. Ibid.

Iv yo' pay us a visit, we've not the least deawt,
But yo'll all on yo' have a reet splendid blow eawt.
Iv yo' come, yo'll want somewheer to lay deawn yore yeads,
So yo'd better write o'er an' secure a few beds;
An' these deawn at Blackpool, yo' know, are furst-class.
Well, good-bye for th' present. Bring plenty o' brass.[118]

Holiday memories and reviews of specific destinations were an additional medium for the circulation of emotion knowledge. Deserving of particular mention is one book that deliberately instrumentalized personal holiday accounts in order to instruct and educate working-class women. The first part of a series of books titled *'Work and Leisure'—Manuals to The Employment and Recreation for Women Workers* was published in 1887. This volume was dedicated specifically to the question of 'Where to Spend a Holiday', and is in fact a compilation of anonymous holiday reviews allegedly written by women for women.

Like other holiday advice texts, the collection reinforced the notion of the holiday as a necessity, pointing to the importance of both physical and emotional well-being. The author of the introduction stressed the need of the working woman to have at least one 'thorough holiday annually, of not less than four weeks', and noted the instructive purpose of the manual.[119] Many of the individual reviews in the book reiterated the need of experienced advice when choosing the right destination and acquiring as much happiness and health from the holiday. '[T]here are many places in England where a holiday may be pleasantly spent at moderate expense', noted one of the writers, 'but [it is] sometimes a little difficult to fix upon such a place'. Hence, '[a] few practical hints may therefore not be amiss of some of those where I... have passed some happy and health giving holidays'.[120]

An amalgam of personal accounts, practical information for the holidaymaker, and advertisements for seaside accommodations, it is hard to place this book into one single category. Moreover, not only did health considerations appear in many of the individual chapters—the content of the introduction is identical to that used in the medical literature:

[118] 'Leave your business and books, dress yourself up in your best / And come watch the sun as he sinks into the west... Is it wise on you to sacrifice pleasure and health / For that nick-named, perishing bauble called wealth? / Nothing of the sort; for you'll never be happy, you'll find / Unless you've good health and a contented mind / Well, we think we've said plenty for one time at least—We've invited you over to a very rich feast / If you pay us a visit, we've not the least doubt / But you'll all on you have a right splendid blow out / If you come, you'll want somewhere to lay down your heads / So you'd better write over and secure a few beds / And these down at Blackpool, you know, are first-class / Well, good-bye for the present. Bring plenty of money.' Ibid.

[119] Anon., *'Work and Leisure' Manuals to The Employment and Recreation for Women Workers. Part I: Where to Spend a Holiday* (London: Hatchards, 1887), iv.

[120] Ibid. 19.

> [O]ne of the most important lessons women have to learn, in view of the more or less professional duties that they are undertaking, is, that not only a change of occupation but a change of *scene* (original), and, as much as possible, of thought and actual distraction of mind, is at certain intervals absolutely necessary to recreate their powers and to maintain them in health and vigour.[121]

Accounting for the emotional essence of holidaymaking reveals that treating holidaymakers as mere purchasers of a product or service, and their needs and desires as mere objects of manipulation is inaccurate. The role of holidaymakers in the emotional economy of holidaymaking transcended their evident function as consumers in the narrow sense of the word. Chapter 4 demonstrated that vacationers participated in the negotiation of the value ascribed to emotional experiences at the resort through their assessments of emotional change. These assessments of the diversity of consumers' experiences, demands, and means of satisfying them in turn contributed to the form taken by the product itself. But the argument can be taken one step further: operating both as spectators as well as performers of the holiday at the resort, vacationers' experiences, interpretations, reactions, and bodies were integral components of the commodified experience that they were themselves consuming. In other words, holidaymakers functioned as the consumers, the producers, and the product.

Finally, holidaymakers were producers and consumers not only of the holiday, but also of emotion knowledge about holidaymaking. They consumed and co-produced ontologies of emotional states and experiences as well as the commodity form of these emotions and experiences. The broad popularity of holiday stories, poems, and reviews, and their large circulation, even among those who could not—at least not yet—afford to experience the holiday themselves, suggests that their function was similar to that of advertising.[122] Like advertisements, these texts mediated the cultural definition of the holiday (the commodity) and its sensorial and emotional meaning, thus shaping and guiding the actual experience of the holidaymaker (the consumer).[123] These popular accounts thus represent competing forms of knowledge about or interpretations of holiday practices and experiences within the broader Victorian and Edwardian emotional economy.

[121] Ibid. iii–iv. [122] See also Beetham, 'Lancashire Seaside Publications', 333–4.
[123] Illouz, 'Emotions, Imagination and consumption', 405–6.

Conclusion

The emotional, the moral, and the economic

'The writer of a book on holidays', argued MD Charles Musgrove, is no different than 'the proprietor of a large hotel, who has to cater for visitors with all sorts of palates and powers of digestion.' Therefore, he explained, a doctor must

> be prepared to guide varied types of individuals as to the form of vacation best adapted to their particular needs. It is not sufficient for him to point out the attractions of different sorts of holidays; he must go further than this. The old saying that 'what is one man's meat is another man's poison' applies to holidays as much as it does to diet.[1]

The comparison between the medical expert and the hotel owner is more than a metaphor. The physicians and early psychologists we have encountered in this book were in many ways formulating both a diagnosis and a cure. True, this is usually the case in medicine; however, the cure in this case was not given in a clinic or an operation room, but 'outsourced' to a commercial leisure industry that was in itself related to a long process of social debate. Positioning themselves as the authority on questions of emotion management, by the 1880s physicians were commenting and advising on almost every component of holidaymaking, from the proper work–leisure ratio, to the determination of the appropriate destination, to the architecture, fees, and particular activities at the resort.[2] In light of the influence that medical discourses exerted on diverse actors and institutions related to late-Victorian and Edwardian British holidaymaking, I began this book by proposing that we rethink the depathologization narrative that is still generally accepted in the historiography on holiday resorts, especially in watering places. My argument throughout was that a more accurate narrative should not contend that the therapeutic essence of holidaymaking was replaced by fun and amusement, but that the objective of the holiday changed from physical cure to emotion management, which was framed in therapeutic terms.

Historians of emotions, and especially historians of emotions in nineteenth-century Great Britain, have argued that although presented by contemporaries as

[1] Musgrove, *Holidays and How to Use Them* (1914), 9.
[2] Benger Alaluf and Illouz, 'Emotions in Consumer Studies'.

The emotional economy of holidaymaking: Health, pleasure, and class in Britain, 1870–1918. Yaara Benger Alaluf, Oxford University Press (2021). © Yaara Benger Alaluf. DOI: 10.1093/oso/9780198866152.003.0007

amoral, the category of emotions was always morally charged.[3] Moreover, as Thomas Dixon shows, the semantics of the Victorian term 'moral' were associated both with psychological issues and with ethical ones.[4] This is evident also in the specific case of holidaymaking. Firstly, like other scientific discourses, the medical conceptualization of emotions 'claimed truth'—to use Nikolas Rose's formulation[5]—by implying ethical guidelines for individual conduct, prescribing patients particular behaviours or activities, and producing and circulating medical advice. Secondly, by categorizing social phenomena, such as work, under a taxonomy of healthy versus pathological, physicians staked out an authoritative position for giving counsel on such matters. As I showed in Chapters 1 and 2, assuming this new position enabled the medical community to produce naturalized explanations and justifications for social problems, thus directly influencing political debates. Thirdly, though there was not uniformity of opinion in the medical community on the matter of emotions, in treating the latter as a fundamental component of individual identity and as a tool for ascribing meaning to social experiences, this discourse both presupposed and constructed a particular kind of subject. This subject, as showed in Chapter 5, was increasingly engaged in introspection and in the pursuit of positive emotional experiences. Finally, by linking emotions both to social structures and to allegedly autonomous and self-responsible individuals, the medical conceptualization of overwork and holidays promoted a 'privatized' ethos of oppression. By holding that emotional alleviation was accessible to everyone, the idea of holiday-as-change made workers responsible for their own well-being and preserved the order that created the need for change in the first place.

Nonetheless, the impact of medical views on holidaymaking was by no means unilateral or impervious, nor was medicine the only sphere of knowledge production significant for holiday culture. For instance, Chapter 2 called attention to a complex dynamic of Victorian gendered views, showing that the incorporation of women into the work force in parallel to the emotional pathologization of work led to a more egalitarian approach to holiday legislation. I have also argued that despite of their middle-class bias, the knowledge produced by Victorian and Edwardian scientists facilitated the provision of amusement and idle holidays to lower classes. As we have seen, resort publicity and popular literature sometimes deviated from or undermined the common medical narrative; yet they also contributed to the creation of a self-reflective consumer, who was at the same time increasingly preoccupied with the emotions of others. This in turn generated the affinity between holidaymaking and authenticity. This idea, too, was contested: claims like 'it is only on a holiday that people are quite themselves' and

[3] See, for instance Boddice, *The Science of Sympathy*; Alberti, 'Introduction: Medical History and Emotion Theory'; Dixon, 'Patients and Passions: Languages of Medicine and Emotion, 1789–1850'.

[4] Dixon, 'Patients and Passions: Languages of Medicine and Emotion, 1789–1850', 39, 43.

[5] Rose, *The Psychological Complex*, 7. See also Illouz, *Saving the Modern Soul*, 44.

therefore that a 'man who is easily recognisable on a holiday is not having a holiday at all'[6] reflect a perception of identity and authenticity that contradicts the dominant criticism of leisure consumption expressed by contemporaneous intellectuals and moralists.

Conceptual challenges and opportunities

I chose to use the model of 'emotional economy' precisely in order to account for these parallel bodies of emotion knowledge and to examine their interrelations. In this sense, more than pinning down a causal mechanism, this book explored the convergence of different socio-cultural processes involved in a moment of historical change. My conceptualization of emotional economy was presented in the introduction and applied throughout the preceding chapters. However, its analytical distinction from related concepts is enlightening in itself, and it can help us understand some of the challenges involved in the study of the interrelation between economy, emotions, and morality.[7]

The concept 'emotional economy' is a semantic relative of the more familiar concept 'moral economy'. Although used quite frequently by eighteenth-century writers, moral economy is usually associated with E. P. Thompson's study on the 'The Moral Economy of the English Crowd in the Eighteenth Century'.[8] Following Thompson's reinvigoration of the concept, the formula '_______ economies' was soon applied in the study of diverse social phenomena. We can now find numerous studies of religious economy, spatial economy, cultural economy, and affective economy (all also with the plural 'economies'), to name just a few examples. Actually, a simple inquiry in academic databases shows that in the last fifteen years, the use of the formula '_______ economies' rose by more than 400 per cent compared to its use in the thirty years between Thompson's 1971 article and the beginning of the twenty-first century.[9] However, not unlike other concepts in the humanities and social sciences, it is usually not easy to find a clear definition of moral economy and related terms, and scholars apply it in diverse ways. Indeed, with the spreading use of the formula came critique, and a number of authors have

[6] Ernest Clephan Palmer, 'Holidays to Taste', *The Daily News and Leader*, 12 June 1914/NYCRO/DC/HRG/BPC 1914.

[7] Parts of the following section are based on a paper titled 'Moral Economies, Emotional Economies and Some Terminological Challenges in the Cultural Research of Economy' presented at the conference *New Directions in Theories of Capitalism* at The Van Leer Jerusalem Institute, in July 2016. I would like to thank Edgar Cabanas, Mattan Shachak, and Ori Shwartz for their comments and my colleagues from IMPRS Moral Economies of Modern Societies, and especially Marie Schubenz and Çiçek İlengiz, for endless discussions about moral economies.

[8] Edward P. Thompson, 'The Moral Economy of the English Crowd in the Eighteenth Century', *Past & Present* 50 (1971): 76–136.

[9] I conducted the analysis for the terms 'moral', 'emotional', 'affective', 'religious', 'spatial', and 'cultural' in conjunction with 'economy' and 'economies'.

pointed to some rhetorical, theoretical, and ideological problems in the application of '_______ economies'.[10] In light of the terminological confusion and the growing scepticism towards it, I would like to dedicate the next pages to outlining different approaches to 'emotional economy' and to draw from this discussion some conclusions about research on emotions and economy.

In 'The Moral Economy of the English Crowd in the Eighteenth Century', Thompson used the concept as a way to analyse rioters' motives beyond a simplistic explanation as 'rebellions of the belly'. Rather, he insisted, their motives were 'grounded upon a consistent traditional view of social norms and obligations'. This 'moral economy of the poor' was based on the old feudal social order, which he called the 'moral economy of provision' and which was gradually being replaced by the emerging 'political economy' around the same time. Proponents of the latter, Adam Smith being the most prominent among them, imagined their theory as being 'disinfested of intrusive moral imperatives'.[11] A close reading of Thompson's work suggests that he applies the term 'moral economy' in at least two different ways. In the first one, which I will call 'an economy which is moral', he traces the moral foundations of a specific socio-economic order and the actions regarded as legitimate for reconstituting it in cases of deviation. In the second interpretation, which I will call 'an economy of morals', he uses it to denote the operation of a system of reciprocity and negotiation in which there is a shared understanding of the rights and obligations of different actors.

This dual meaning stems from the polysemy of the term 'economy'.[12] One meaning of 'economy', most frequently used nowadays, refers to specific sciences, ideologies, and systems of production, labour, trade, and consumption. The second meaning is allegedly closer to the Greek etymology of the word *οίκονομία*, which is composed of the words 'house' and 'management'. *Oίκονομία*, or oeconomy, was historically used abstractly to refer to the knowledge required for a proper management of the household, or more generally for the efficient management of

[10] See, for instance Jaime Palomera and Theodora Vetta, 'Moral Economy: Rethinking a Radical Concept', *Anthropological Theory* 16, no. 4 (2016): 413–32; Johanna Siméant, 'Three Bodies of Moral Economy: The Diffusion of a Concept', *Journal of Global Ethics* 11, no. 2 (2015): 163–75; Chris Gibson and Lily Kong, 'Cultural Economy: A Critical Review', *Progress in Human Geography* 29, no. 5 (2005): 541–61.

[11] Thompson, 'The Moral Economy of the English Crowd in the Eighteenth Century', 77–9, 89–90, 132.

[12] I refer here to contemporary predominant Western meanings of 'economy' and to the ways in which its etymology is generally used nowadays, and not to the historical meanings of *οίκονομία* or to various non-Western traditions of economy. For an insightful analysis of the differences between 'what we call "the ancient economy"' and 'what the ancients called "the oikonomia"' see Dotan Leshem, 'Oikonomia Redefined' *Journal of the History of Economic Thought* 35, no. 1 (2013): 43–61. For an exhaustive philological and philosophical history of oikonomia, see Dotan Leshem, *The Origins of Neoliberalism: Modeling the Economy from Jesus to Foucault* (New York: Columbia University Press, 2016).

resources.[13] It is important to note then, as Norbert Götz emphasizes, that in classic philosophy, the concept 'moral economy' would have been redundant, and it thus makes sense that the term only appeared in the eighteenth century, when morality was gradually detached from the notion of economy.[14]

This polysemy of the term 'economy' is the source of most of the ambiguity and confusion surrounding the use of the '_______ economies' formula, which, following Thompson, continued to carry with it the two meanings noted above. Hence, on the one hand, we have 'an economy which is moral', which is to say the study of how economy is influenced by moral dispositions and how these are shaped by economic pressures. Both Thompson and James Scott, in his work on Southeast Asian peasants,[15] focused on cases of economic unrest to analyse conflicts of subsistence and the interplay of paternalistic authority, popular values, and mutual expectations. Indeed, both can be criticized for reaffirming a Polanyi-influenced opposition between peasant economies and capitalist ones. Still, keeping in mind the danger of nostalgic romanticism, this approach is useful for emphasizing processes of legitimation and de-legitimation of particular economic orders, as its core argument is that revolts do not just express some kind of survival instinct, but reflect a sense of morality. This use of moral economy rejects the notion that self-interest governs human behaviour and challenges the idea of a market or an economy autonomous from the social world and its related conceptions of right and wrong.

On the other hand, we have 'an economy of morals', which is to say, the theorization of morality as structured in a dynamic system of values, norms, and cultures. The most prominent representatives of this approach are historian Lorraine Daston and anthropologist Didier Fassin. In her article 'The Moral Economy of Science', published in 1995, Daston chooses the concept 'moral economies' in order to describe 'a web of affect-saturated values that stand and function in well-defined relationship to one another' and therefore to analyse the social negotiation of complex normative codes.[16] Fassin also uses the term 'economy' to account for a system of negotiation between contested values, defining moral economy as a system that consists in '[la] production, la répartition, la circulation et l'utilisation des émotions et des valeurs, des normes et des obligations'.[17] But in contrast to Daston's use, here 'economy' does not simply stand for

[13] Leshem, 'Oikonomia Redefined', 44; Norbert Götz, '"Moral Economy": Its Conceptual History and Analytical Prospects', *Journal of Global Ethics* 11, no. 2 (2015): 147–62, here 148–9.

[14] Götz, 'Moral Economy', 148–9.

[15] James C. Scott, *The Moral Economy of the Peasant: Rebellion and Subsistence in Southeast Asia* (New Haven: Yale University Press, 1976).

[16] Lorraine Daston, 'The Moral Economy of Science', *Osiris* 10, no. 1 (1995): 2–24, here 4; see also Boddice, *The Science of Sympathy*, 13.

[17] 'The production, distribution, circulation and use of emotions and values, norms and obligations' (my translation). Didier Fassin, 'Les économies morales revisitées', *Annales. Histoire, Sciences Sociales* 64, no. 6 (2009): 1237–66, here 1257; see also Didier Fassin, 'Compassion and Repression: The Moral Economy of Immigration Policies in France', *Cultural Anthropology* 20, no. 3 (2005): 362–87.

'order', but returns to the original concept of oeconomy in order to elucidate the relation between moral values and material resources.[18]

In recent decades, scholars have adopted the '_______ economies' formula in the growing body of research on emotions, while generally retaining the division outlined above. Although they do not always apply the concept 'emotional/affective economies' in a literal sense, under the first interpretation we can situate some of the central works on capitalism and emotions, which point to the various ways in which late capitalism functions as 'an economy which is emotional'. These works include: semiotic approaches that emphasize the meaning that commodities carry beyond their material or functional value;[19] the different accounts of affective or emotional labour that point to the production, organization, and manipulation of affect for the needs of the capitalist system;[20] media scholar Henry Jenkins' concept 'affective economics', which accounts for affective components in consumer decision-making processes as well as for the ways in which the active emotional engagement of the consumer (e.g. hitting the 'like' button or audience voting in reality TV) accumulates into data that is then turned to profit;[21] and last but not least, Eva Illouz's use of the term 'emotional capitalism' with the aim of explaining the evolution of a culture in which economy and emotions increasingly converge.[22] All these scholars have pointed to the incorporation of emotions in the economic sphere and, moreover, to the mutual constitution of emotional and economic discourses and practices. These approaches have contributed immensely to the study of late capitalism and have helped to create an understanding of economic relations as being more than rational choices made by self-interested individuals. In this sense, they also serve as a constructive deviation from Thompson and Scott's moral economy, which sought to attribute to the masses a collective rational decision-making process.

Other scholars doing research on emotions have used the term 'economy' to describe a dynamic system, or 'an economy of emotions'. Rob Boddice follows Daston's concept of moral economies in order to account for parallel 'modes of thinking, feeling, and acting within specific "communities"'.[23] For the aims of his research—understanding changes in ideas and practices of both feelings and

[18] See also Siméant, 'Three Bodies of Moral Economy'; Götz, '"Moral Economy"'.

[19] See for instance Arthur Asa Berger, *The Objects of Affection: Semiotics and Consumer Culture* (New York: Palgrave Macmillan, 2010); Viviana Zelizer, *The Purchase of Intimacy* (Princeton: Princeton University Press, 2005); Jean Baudrillard, *The Consumer Society: Myths and Structures* (London: Sage, 1998); Arjun Appadurai (ed.), *The Social Life of Things* (Cambridge: Cambridge University Press, 1986).

[20] Antonio Negri and Michael Hardt, 'Value and Affect', *boundary 2* 26, no. 2 (1999): 77–88; Arlie Russell Hochschild, *The Managed Heart: Commercialization of Human Feeling* (Berkeley: University of California Press, 1983).

[21] Henry Jenkins, *Convergence Culture: Where Old and New Media Collide* (New York: New York University Press, 2006).

[22] Illouz, *Cold Intimacies: The Making of Emotional Capitalism.*

[23] Boddice, *The Science of Sympathy*, 13.

morality among Victorian scientists—Daston's understanding of 'economy' is well suited and effective. Putting more emphasis on the notion of 'value' rather than on 'system', Ute Frevert's work with the concept 'emotional economies' implies a different meaning to the term 'economy'. Frevert uses emotional economies as a framework for reading historical changes as responses to emotional demands, and to understand when and why some emotions gain higher value than others.[24] Yet a different notion of economy can be found in the work of Sara Ahmed. In 2004, Ahmed coined the concept 'affective economies' in order to theorize the ways in which the circulation of emotions in society materializes subjects, bodies, groups, and social relations. Ahmed understands emotions as being social, material, and psychological. In this sense, she treats affects or emotions not as something that reside in the subject or that are 'about' their object: in Ahmed's own words, in 'affective economies, emotions *do things* (original), and they align individuals with communities—or bodily space with social space—through the very intensity of their attachments'.[25] An affective economy is thus a continuous process of circulation, in the course of which the very surfaces of individual and communal bodies are produced. In other words, and in contrast to Fassin, Ahmed approaches emotions as phenomena that are not only produced in society, but that are constitutive of the social itself.

In line with the empirical analysis of holidaymaking undertaken in this book, I would like to supplement the discussion of these conceptual and rhetorical issues by raising some broader concerns. In many cases, the use of the '_______ economies' formula seems to allow scholars to avoid the harder theoretical work needed to get a firm grasp on the complex relations that the concept entails. Hence, the first point that I want to make can be summarized in simple terms: it's complicated. Above, I offered a way to make sense of the different uses of 'moral/emotional economies' through a rather simple distinction between two paths: one tackles the reciprocal influence between economy and emotions/morality and the second deals with processes of production and circulation of moralities and emotions. But what happens if these two paths converge? In truth, any 'economy which is moral/emotional' always has to correlate with a particular 'economy of morals/emotions'. Thus, I contended that cultures of holidaymaking depend on two processes: emotions are produced as meaningful, desirable, and manageable objects *and* as experiences that can be deliberately consumed in commercialized spaces.[26]

Furthermore, these processes are not necessarily independent of one another. Consider, for example, the medical literature on emotions analysed throughout the book: this literature objectified emotions by formulating culturally distinct conceptions of emotions and emotion management. But at the same time, this

[24] Frevert, *Emotions in History*, 212. [25] Ahmed, 'Affective Economies', 119.

[26] See also Benger Alaluf and Illouz, 'Emotions in Consumer Studies'.

literature was itself commodified: first, it enhanced the prestige of a specific genre of holiday advice literature, and second, it provided foundational justifications for creating new positions for physicians within the resort industry.[27] Chapter 2 showed that the emerging medicalized epistemology of emotion also influenced the labour sphere by providing justification for the reduction of working time and the provision of holidays in more and more professions. Likewise, Chapters 4 and 5 demonstrated that the reception and reaction of individuals to the (commercialized) medical advice and to the (consumerist) holiday experience reproduced and modified the cultural meanings and repertoires of emotions as well as the resort industry in which some of these were being experienced, through a variety of material, spatial and symbolic mechanisms. Furthermore, a central argument developed in this book was that the practice of holidaymaking was performative, meaning that it generated emotions in consumers. From this follows that the emotional economy of holidaymaking also involves the production of the subjects and institutions that constitute it. I hope to have shown that sensitive historical analysis can disentangle some of these dynamics.

Historicization is crucial here, and this is my second point. The use of emotional economies as a concept must not assume the existence of a-priori, fixed meanings of either economic systems or emotion knowledge. As an example, in her use of 'affective economies', Ahmed takes the idea of the 'lacking subject' as a given, thus extrapolating the Lacanian theory of emotions from the psychic field to the social.[28] Instead of exposing the historical specificity of this emotion knowledge, Ahmed overlooks the role of psychoanalysis in producing particular perceptions of emotions, and in doing so runs the risk of converting the psychoanalytic subject into a universal type. Saying that social bodies are produced through the circulation of emotions is not enough if we fail to account for the historical specificity of the emotional economy involved.[29]

As the works cited above evidence, the research on emotional economies has been mostly limited to the study of late capitalism and neoliberal societies. In many cases, the concept seems to be used for theoretical goals rather than for empirical study. However, the concept of 'emotional economies' can still be a useful tool for historical research, helping us to trace specific historical processes, as did Thompson's original use of moral economies. This, in turn, can facilitate

[27] For a similar argument about contemporary psy-industries (e.g. psychopharmacology, coaching, emotion management workshops), see Mattan Shachak, '(Ex)changing Feelings: On the Commodification of Emotions in Psychotherapy', in Eva Illouz (ed.), *Emotions as Commodities: Capitalism, Consumption and Authenticity* (London: Routledge, 2017).

[28] See Ahmed, 'Affective Economies', 120.

[29] On the need to historicize emotions as an object of inquiry and on the risks of psychological essentialization of emotions, see Boddice, 'The History of Emotions: Past, Present, Future'; on the importance of historical analysis in the study of emotions and consumption in particular see Benger Alaluf and Illouz, 'Emotions in Consumer Studies'; Eva Illouz, 'Toward a Post-Normative Critique of Emotional Authenticity', in Eva Illouz (ed.), *Emotions as Commodities: Capitalism, Consumption and Authenticity* (London: Routledge, 2017).

efforts to historicize established arguments in other disciplines. For example, the analysis of Victorian and Edwardian holidaymaking in Britain suggests that 'emotional capitalism'[30] and the 'experience economy'[31] have their origins in the late nineteenth century.

Significant too is the insight that the language we use is historically and culturally specific. As we have seen, most of the confusion in the different approaches to moral and emotional economies derives from their different contemporary understandings of the term 'economy'. The classic meaning of 'oeconomy' has been more or less lost in present Anglophone culture, and the contemporary use of the term 'economy' is so rooted in it that returning to the term's older meaning presents serious challenges. Nevertheless, I believe that doing so is an important task, both for the sake of conceptual coherency and as a way to challenge the contemporary hegemonic definition of the term by reclaiming it to the humanities. An unreflective use of the term 'economy' as an analytical tool brings along its own dangers. In a way, Thompson was wrong to conjoin the terms 'morality' and 'economy', because in doing so he implied that an economy without morality was possible, an approach that conforms with the neoclassical ideology. Recent uses of 'moral/emotional economies' that treat the term 'economy' as a proxy for 'system' or 'order' operated by some kind of invisible hand occasionally fall into a similar trap. The problem is that in doing so, the concept 'economy' becomes a presentism that might dissolve into the general discourse of neoliberalism, thus accomplishing little more than contributing to the ongoing economization of subjects and social relations.

Besides the conceptual reflection, this risk can be eluded through a methodology that acknowledges actants on all levels. A major advantage of the model of 'emotional economies' informing this book's analysis is that it accounts for the dynamics of negotiation, which is not necessarily a unilateral top-down process, just as it does not favour institutions or discourses over individuals. The model gives the historian a tool to elucidate how actors' practices and choices as well as social norms generate historical change.[32] This brings me to my third and last point: a constructive use of the 'moral/emotional economies' model in each of its interpretations must take agency seriously in order not to become 'a causal shortcut'[33] that overshadows actual historical processes and those who are involved in them. By incorporating personal accounts of holidaymakers—their own reflections on the meaning of their actions, choices, and emotions—this book made a contribution to the growing body of research that challenges the long

[30] Illouz, *Cold Intimacies: The Making of Emotional Capitalism.*

[31] Pine and Gilmore, *Experience Economy.*

[32] See also Alan Warde, 'After Taste: Culture, Consumption and Theories of Practice', *Journal of Consumer Culture* 14, no. 3 (2014): 279–303.

[33] See also Siméant, 'Three Bodies of Moral Economy', 170–1.

tradition of assuming that the rational and the emotional stand in stark opposition to one another.[34]

Homo economicus on the beach

These insights compel us to revise some assumptions commonly made in theorizations of holidaymaking, tourism, and leisure in general, the first and most obvious being: emotions matter. Historians and sociologists of tourism have recognized this fact, but once and again they have dismissed it. Thus, in their influential article 'The trouble with tourism and travel theory?' Adrian Franklin and Mike Crang observe that 'it is not just bodily pleasure that have often been downplayed but pleasure *tout court*' and admit that '[r]eading the literature it would be hard to get a real sense that perhaps the central feature of tourism is pleasure, fun and enjoyment'. They acknowledge that researchers of tourism 'need to be able to say tourism matters because it is enjoyable, not in spite of it', yet shrug it off by saying that they 'lack a language that can speak to the enjoyment and pleasure of tourism'.[35] John K. Walton, as well, asserts that leisure and holiday industries 'tend to remain hidden from the historian's gaze because they sell experiences, even dreams, intangible products which generate neither the convincing statistics which economists require... nor the official files and reports constructed by and for officials and ministries, which constitute the usual sources for historians'.[36] Indeed, as he notes, the historiography on leisure, sport, tourism and holidaymaking developed astonishingly in the past decades, but the dreams, experiences and intangible products remain obscure. Last but not least, in his careful and important discussion about the challenges of formulating a definition of leisure, Peter Borsay notes:

> What is often meant [in some definitions of leisure] is 'freedom from' rather than 'freedom to', and even in the latter case it is often unclear what sort of experience is being envisaged; freedom to do what? The pursuit of pleasure might be the answer. However, quite apart from the problem of accommodating recreations which embrace elements of pain and punishment, and there are surprisingly many of these, the notion remains too intangible to be satisfying.[37]

Borsay then presents his own definition of leisure and systematically implements it throughout the book. But in the very last paragraph he frames leisure as a

[34] See for instance Illouz (ed.), *Emotions as Commodities*; Illouz, *Cold Intimacies: The Making of Emotional Capitalism*; Ute Frevert, 'Passions, Preferences, and Animal Spirits: How Does Homo Oeconomicus Cope with Emotions?', in Frank Biess and Daniel M. Gross (eds), *Science and Emotions after 1945: A Transatlantic Perspective* (Chicago: University of Chicago Press, 2014).

[35] Adrian Franklin and Mike Crang, 'The Trouble with Tourism and Travel Theory?' *Tourist Studies* 1, no. 1 (2001): 5–22, here 14.

[36] Walton, 'The History of British Spa Resorts', 140.

[37] Borsay, *A History of Leisure*, 6.

response to the pressures of social structures 'from which occasional release is a psychological imperative'.[38] It was the intention of the present study to show that the emotional, whether related to the feeling of freedom, pleasure, boredom, disappointment or fear, can be and should be historically analysed in order to get to the bottom of the leisure experience. The history of emotions leaves much less room for excuses for such dismissive claims and no place for a-historical assumptions regarding 'the psychological'.

Taking emotions seriously in the research of holidaymaking and tourism implies questioning some views concerning the holidaymaker as consumer. Leisure is not, as Borsay contends, 'surely something undertaken when a person is not engaged in economic activity';[39] nor can we say that leisure time, once commodified, necessarily 'becomes another routinized, packaged, commodity, thereby failing to be anything like a carefree, relaxed, alternative to work'.[40] It should be clear at this point that I see commercial leisure activities as both economic and emotional. Moreover, drawing on the work of Eva Illouz and her colleagues and working against the typical critique of the Frankfurt School,[41] I claimed that in an emotional economy, products should not be seen as 'an arbitrary choice forced on consumers', but rather as 'an outcome of broader cultural processes'[42] that involve conscious consumers. As I showed in Chapter 5, the authentic experience of emotions was linked to the performative character of holidaymaking, which merges the emotional and the economic. In a culture in which the ideal of authenticity is increasingly linked to (what are perceived as) personal emotional experiences, the question should not be whether this experience is authentic or not; rather, taking the statements and writings of self-reflective actors seriously, we should trace the emotional economy that makes some experiences 'feel authentic'.[43] Additionally, the emotional economy approach, as was exemplified mainly in Chapter 4, also calls to consider the emotional experience of holiday as co-produced by various actors and objects, including resort managers, medical experts, atmospheres, artefacts, and the vacationers themselves.[44] Without downplaying important issues concerning labour and exploitation in the tourism industry, this puts into question interpretations describing tourism as consisting of 'two categories of people: those who demand, and those who supply'.[45]

[38] Ibid. 227. [39] Ibid. 40.

[40] Watson and Kopachevsky, 'Interpretations of Tourism as Commodity', 645.

[41] Most notably: Marcuse, *One-Dimensional Man*.

[42] Illouz and Benger, 'Emotions and Consumption', 268.

[43] Illouz, ed., *Emotions as Commodities*.

[44] Illouz, 'Introduction: Emodities or the Making of Emotional Commodities'; Ramirez, 'Value co-production'; although I agree with Illouz and Ramirez that the dynamic of co-production are a central feature of neo-liberal societies, I argue that manifestations of it can be traced to the end of the nineteenth century.

[45] Watson and Kopachevsky, 'Interpretations of Tourism as Commodity', 653.

Taking emotions seriously also calls for a reappraisal of the meanings of rationality, both for individuals as for collectives. The emotional economy of leisure consumption and the possibility of attaining certain emotional experiences through consumerist practices—a potential that was recognized, as we have seen, by contemporaries—suggests that in some cases, emotions are closely linked to public, commercial spaces. Against the Habermasian emphasis on rational communication as the fundamental constitutive element of the public sphere[46] and historians of Victorian and Edwardian leisure cultures who frequently reiterate the contemporaneous middle-class notion of 'rational recreation',[47] the sources analysed in this book unravel the significant role of emotion knowledge and emotional habitus in the formation of leisure spaces, collective experiences, and political communities.

This emotional economy also calls into question the view of *homo economicus* as essentially rational. Indeed, from the 1980s onwards, this has been a central focus of much of the scholarship on consumer culture. An emotional economies approach, however, aims not only at pointing to the co-existence of rational utilitarian action and semiotics, but seeks to trace the processes through which particular economic and emotional repertoires came to be interrelated. Victorian and Edwardian holidaymakers had access to different sources of information which they considered in order to make well-informed decisions about the destination and the activities of their holiday, and they made the decision they believed would result in the highest personal benefit. As we have seen, some also retrospectively evaluated the value of their holiday experience. This did not make them less emotional. On a micro level, they achieved a particular emotional experience (and if you recall Ab-o'th'-Yate and others, they also engaged in intense emotional work[48]). On a macro level, the emotional economy of holidaymaking, which included medical knowledge, marketing efforts, and political discourses, helped to enhance an *ethos* of emotional self-management, which made emotional subjectivity consistent with goal-oriented rationality.

Taking emotions seriously can help to sharpen some theoretical and conceptual approaches to leisure, tourism and holidaymaking, and to guide historical research. Many scholars have addressed some paradoxes of holidaymaking because of the challenges they pose to the perception of leisure in general and of

[46] Jürgen Habermas, *The Structural Transformation of the Public Sphere: An Inquiry into a Category of Bourgeois Society*, trans. Thomas Burger (Cambridge: MIT Press, 1991). See also Karl Spracklen, *The Meaning and Purpose of Leisure: Habermas and Leisure at the End of Modernity* (London: Palgrave Macmillan, 2009).

[47] Most notably Bailey, *Rational Recreation*.

[48] Note that Arlie Hochschild's influential concept of 'emotional labor' focused on the transformation of (workers') emotions to (employers') economic profit, without accounting for similar patterns in the sphere of consumption, i.e. the case of consumers who pay in order to alter their emotional state for their own individual benefit. Hochschild, *The Managed Heart*; see also Benger Alaluf and Illouz, 'Emotions in Consumer Studies'; Benger Alaluf, 'Exploring the Production of Relaxation in Club Med Seaside Resorts', 47, 49–50.

the British resort as a carnivalesque space.[49] Recent literature is questioning the applicability of the concept, pointing to the many ways in which the holiday was in fact far from being an experience of exotic individual freedom, but was rather regulated, collective, familiar and even predictable.[50] This ongoing debate could benefit from taking into consideration the emotional prism, for example by understanding the core of the carnivalesque to be an emotional experience. If we take the 'difference' or the 'other' that constitutes the holiday experience to be not an opposition of objects, places or ideas (e.g. modernity and urbanism on the one hand, tradition and nature on the other), but as contingent on the emotions evoked by them, we can understand the holiday not as carnival, but as an embodied emotional state. We should therefore explore the history of 'the "holiday mood"' rather than that of 'the carnival spirit',[51] and of tourism practices as a 'controlled de-control of the emotions'.[52]

Finally, taking emotions seriously charts new directions for future research. This book unfolded the configuration of emotional and economic ideas, experiences, and practices reflected in the holiday culture of a particular time and place. I showed that holidaymaking in late-nineteenth- and early-twentieth-century British resorts was the outcome of broader processes, including the emotionalization of nature, the pathologization of work and the worker, the democratization of leisure, and the commercialization of the local government. Extending the time frame of an emotional economies analysis into the twentieth century might shed new light on some unsolved questions, such as the faith of the British spa resort industry, or the emotional economy of the Edwardian amusement park or the British holiday camp.[53] Moreover, some of these processes still affect emotional economies of holidaymaking today. However, in contrast to the history of tourism on the one hand and the history of the emotionalization of consumers on the other, the link between the two—the emotional history of holidaymaking—remains far less studied. This is particularly curious considering the central place of holidaymaking in current societies, especially in relation to stress and

[49] See for instance Rojek, *Capitalism and Leisure Theory*; Webb, 'Bakhtin at the Seaside'; Borsay, *A History of Leisure*, 79, 175; Urry, *The Tourist Gaze*, 29; Shields, 'Liminality and the Carnivalesque at Brighton', Walvin, *Beside the Seaside*, 63.

[50] Borsay, 'Visualising the Seaside, c. 1750–1914', 197; Kane, 'Edwardian Amusement Parks: The Pleasure Garden Reborn?', 240–1.

[51] Kane, 'Edwardian Amusement Parks: The Pleasure Garden Reborn?', 241.

[52] Mike Featherstone, *Consumer Culture and Postmodernism* (London: Sage, 1991), cited in Edensor, 'Performing Tourism, Staging Tourism', 78–9.

[53] Jane Adam's work has shed significant light on the role played by the state in the decline of the British spa. It remains unclear what other forces influenced this process or what exactly differentiates between the traditional spa culture and the contemporary spa industry in Britain, Adams, *Healing with Water*. Josephine Kane's research in the last few years is remarkable, and indicates that there is room for more work on the relation between spaces of consumption and emotional experience, Josephine Kane, *The Architecture of Pleasure: British Amusement Parks 1900–1939* (London: Routledge, 2016); Kane, 'Edwardian Amusement Parks: The Pleasure Garden Reborn?'.

burnout.[54] Today, the link between work, emotional well-being and holidaymaking seems commonsensical, and is supported by popular media and advertisers as well as by quackery and scientific research. Studies on the merits and advantages of holidaymaking are regularly undertaken in diverse academic fields, from public health, psychology, and happiness studies,[55] to economics, marketing, and business management.[56,57] In some cases, these works address both the interests of the holidaymaker and the industry, offering 'recommendations for the success of tourism businesses as well as the enhancement of tourists' senses of well-being'.[58] Similarly to Victorian and Edwardian texts, today's medical literature also emphasizes the responsibility of the patients for their own emotional experience, while doctors position themselves as professional advisers on how to do so. As the authors of one article note: although '[l]eisure time activities do not come easily to all of us', they 'could be encouraged in our patients'.[59] Health practitioners do not only heal; they also produce consumers.

And like back then, many professional takes on holidaymaking make their way to the popular media. Online magazines point to studies that prove that 'going on holiday is good for you', that holidays 'alleviate symptoms of depression', 'evoke happy memories', 'help you to manage stress, improve sleep patterns, reduce blood pressure, strengthen relationships, live longer... and they can even work wonders for the libido... Still sceptical? You obviously need a holiday.'[60] There are also numerous guidelines and quizzes that will help you find out if you need a

[54] Some of the ideas developed in the following appeared in Benger Alaluf and Illouz, 'Emotions in Consumer Studies'.

[55] See for instance Jessica De Bloom, Sabine A.E. Geurts, and Michiel A.J. Kompier, 'Vacation (After-)Effects on Employee Health and Well-Being, and the Role of Vacation Activities, Experiences and Sleep', *Journal of Happiness Studies* 14, no. 2 (2013): 613–33; Charlotte Fritz and Sabine Sonnentag, 'Recovery, Well-Being, and Performance-Related Outcomes: The Role of Workload and Vacation Experiences', *Journal of Applied Psychology* 91, no. 4 (2006): 936–45; Linda L. Caldwell, 'Leisure and Health: Why is Leisure Therapeutic?' *British Journal of Guidance & Counselling* 33, no. 1 (2005): 7–26; Mina Westman and Dalia Etzion, 'The Impact of Vacation and Job Stress on Burnout and Absenteeism', *Psychology & Health* 16, no. 5 (2001): 595–606.

[56] See for instance Cornelia Voigt, Graham Brown, and Gary Howat. 'Wellness Tourists: In Search of Transformation', *Tourism Review* 66, no. 1/2 (2011): 16–30; Melanie Smith and László Puczkó, *Health and Wellness Tourism* (Oxon: Routledge, 2009); Hansruedi Mueller and Eveline Lanz Kaufmann, 'Wellness Tourism: Market Analysis of a Special Health Tourism Segment and Implications for the Hotel Industry', *Journal of Vacation Marketing* 7, no. 1 (2001): 5–17.

[57] Benger Alaluf and Illouz, 'Emotions in Consumer Studies'.

[58] Chun-Chu Chen, Wei-Jue Huang, and James F. Petrick, 'Holiday Recovery Experiences, Tourism Satisfaction and Life Satisfaction—Is There a Relationship?' *Tourism Management* 53 (2016): 140–7, here 140.

[59] Linda Buettner, Mona Shattell, and Madeleine Reber, 'Working Hard to Relax: Improving Engagement in Leisure Time Activities for a Healthier Work-Life Balance', *Issues in Mental Health Nursing* 32, no. 4 (2011): 269–70.

[60] Adrian Bridge, 'Holidays Are Good for You—It's Official', *The Telegraph*, 31 January 2013, accessed 12 February 2019, http://www.telegraph.co.uk/travel/news/Holidays-are-good-for-you-its-official/; Naomi Colema, 'Why Holidays are Healthy', *Daily Mail*, accessed 12 February 2019, http://www.dailymail.co.uk/health/article-60094/Why-holidays-healthy.html.

holiday, 'really need a holiday', or rather 'really, really need a holiday'.[61] Reading these texts, one learns for example that 'irrational anger', 'lack of enthusiasm', and 'public crying' are some of the signs that you need a holiday.[62] The *Business Insider* adds: 'You're starting to make more mistakes at work', 'you feel your work spirit flagging, and can't remember what it is that made you love your job', and '[i]f your entire identity is wrapped into your work—so much so that you've forgotten who the real you is—then it's time for a vacation'.[63]

Once you book that vacation, you can turn to any app shop, where you will find dozens of 'holiday countdowns' with which 'you can count down the days, hours, minutes and seconds until your vacation', apps that will help you to '[b]e inspired, build excitement and get into the holiday spirit'.[64] While on holiday, you might choose to upload a picture and tag it, perhaps with explicit emotional terms, or write in an online review explaining how disappointing your stay at that particular resort was.[65] This information will be immediately turned into data. This data will be used by advertisers, travel agencies, and resorts to adjust their product to trends in consumers' emotional demand. This is an emotional economy of holidaymaking.

[61] Brid O'Connell, 'Quiz: How Much Do You Need a Holiday?' *Thomas Cook Airlines*, 10 February 2017, accessed 12 February 2019, https://blog.thomascookairlines.com/quiz-much-need-holiday/; Maria Masters, '10 Signs You Really Need a Vacation', *Health*, 31 July 2015, accessed 12 February 2019, http://www.health.com/health/gallery/0,20936703,00.html#when-to-take-a-day-off-0; Candace Lowry, 'How Badly Do You Need A Vacation?' *BuzzFeed*, 21 August 2014, accessed 12 February 2019, https://www.buzzfeed.com/candacelowry/how-badly-do-you-need-a-vacation?utm_term=.imp PZdMXX#.xfEKQAlWW; Alison Lynch, '17 Signs You Really, Really Need a Holiday', *Metro*, 23 August 2014, accessed 12 February 2019, http://metro.co.uk/2014/08/23/16-signs-you-really-really-need-a-holiday-4842512/.

[62] Amy Baker, '7 Glaring Signs You Need a Holiday', *Metro*, 16 June 2017, http://metro.co.uk/2017/06/16/7-glaring-signs-you-need-a-holiday-6687106/.

[63] Jacquelyn Smith, '12 Signs You Desperately Need a Vacation from Work', *Business Insider*, 2 January 2016, accessed 12 February 2019, http://www.businessinsider.com/signs-you-need-a-vacation-2016-1?IR=T/#1-you-begin-to-look-like-your-passport-photo-1.

[64] 'Holiday & Vacation Countdown Timer. Developer: Pro Sellers World LLC', *iTunes*, 7 March 2017, accessed 12 February 2019, https://itunes.apple.com/de/app/holiday-vacation-countdown-timer-event-widget/id1034678674?l=en&mt=8; 'Holiday Countdown. Developer: Holiday Extras Ltd', *GooglePlay*, 16 January 2017, accessed 12 February 2019, https://itunes.apple.com/de/app/holiday-vacation-countdown-timer-event-widget/id1034678674?l=en&mt=8.

[65] Benger Alaluf and Illouz, 'Emotions in Consumer Studies'; Benger Alaluf, 'Exploring the Production of Relaxation in Club Med Seaside Resorts'.

Bibliography

Primary sources

Archival sources

Blackpool History Centre (BLP)

Laycock, Samuel. Poems 1871–1908/LT51 Pamphlet

The British Library (BL)

Add. MSS. Additional Manuscripts

General Reference Collection YK.1992.a.5634 [Miscellaneous publications, not catalogued separately. London: Great Northern Railway, 1912–1922].

Harrogate Library, Local Studies Collection (HRG)

PH/615.49 Material relating to baths

H/AAA Harrogate Library, local studies collection

Historical Archive on Tourism, The Technical University of Berlin (HAT)

618/02/906/GRE

The North Yorkshire County Record Office (NYCRO)

DC/HRG Harrogate Borough Records Collection

DC/HRG II 1/5 1875–1882

DC/HRG II 1–5 1891–1896. Wells Baths etc. Committee

DC/HRG II 1/5/1–15 1906–1918. Wells, Baths, Establishment and Advertising Committee

DC/HRG/Book of Press Cuttings, 1911–1912

The Daily Graphic, 20 July 1912

The Daily Mail, 22 August 1912

The Gentlewoman, 12 April 1912; 19 November 1912

The Times, 4 July 1912

Town & Country, 6 May 1912

The World, 12 July 1912

The Yorkshire Evening Post, 15 September 1912

DC/HRG/Book of Press Cuttings, 1914

The Daily Chronicle, 14 May 1914

Daily News & Leader, 10 April 1914; 12 June 1914

The Gentlewoman, 4 July 1914

North Mail, 22 June 1914

The Pioneer, 24 July 1914

Reuter's Bulletin, Port Said, n.d.

Vanity Fair, 25 June 1914

The Yorkshire Evening Post, 29 September 1911; 30 September 1911; 12 October 1911

DC/RIC Ripon Corporation Records

DC/RIC II 1/5/9 1896–1900. Wells, Baths, Establishment and Advertising Committee

DC/RIC II 1/5/10 1900–1906. Wells, Baths, Establishment and Entertainments Committee
DC/SCB Scarborough Borough Records Collection
DC/SCB II 1/11/2. Scarborough Corporation Minutes, Pleasure Boats and Bathing Committee, 1892–
DC/SCB II 1/13/3. Scarborough Corporation Minutes, Parks, Advertising and Entertainments Committee 1909–1911
DC/SCB II 1/13/1. Scarborough Corporation Minutes, Public Walks and Pleasure Grounds Committee 1890–1903

Scarborough Library, Local Studies Collection (SCB)

Trades Union Congress Library Collections at London Metropolitan University (TUC)

HD 5106 Hours of labour including overtime, shift work, sick leave, vacations
HD 9551.5 Mineral industries. Metal trade

Printed primary sources

UK Parliamentary Papers and debates

Commission on the Employment of Children, Young Persons, and Women in Agriculture (1867). First Report of The Commissioners, with Appendix Part I to the First Report. Reports from the Assistant Commissioners and Miscellaneous Evidence. 4068, 4068-I. 1868

Factory and Workshops Acts Commission: Report of the Commissioners Appointed to Inquire into the Working of the Factory and Workshops Acts. Appendix C. C.1443. 1876

Holidays of Lower Division Clerks, Treasury Minute Respecting the Holidays of Lower Division Clerks. C.4118. 1884

Report and Special Report from the Select Committee on the Shop Hours Bill. Together with the Proceedings of the Committee, Minutes of Evidence, Appendix and Index. HC 287. 1892

Departmental Committee on Compensation for Industrial Diseases. Report of the Departmental Committee on Compensation for Industrial Diseases. Report. Cd. 3496. 1907

Holidays. A Bill to Secure One Day's Rest in Seven for All Workpeople; To Make the First Day of May a Bank Holiday throughout the United Kingdom; To Secure a Week's Holiday for Workpeople; and To Secure Payment of Wages for Public Holidays. Bill 242. 1911

Weekly Rest-Day. A Bill to Secure One Day's Rest in Seven for All Workpeople; To Make the First Day of May a Bank Holiday throughout the United Kingdom; To Secure a Week's Holiday for Workpeople; and To Secure Payment of Wages for Public Holidays. Bill 53. 1912

Reports to the Secretary of State for the Home Department on the Shops Act, 1912, as it Applies to Refreshment Premises. By Miss A. M. Ahrons and Mr A. K. Clark-Kennedy, H.M. Inspectors of Factories. Cd. 6579. 1912

Factory and Workshop Act, 1901, Amendment. A Bill to Amend the Factory and Workshop Act, by Reducing the Period of Employment in Cotton Factories to Forty-Eight Hours per Week. Bill 199. 1913

Interim Report on an Investigation of Industrial Fatigue by Physiological Methods, by A. F. Stanley Kent, M.A., D.Sc. (Oxon), Henry Overton Wills Professor of Physiology in the University of Bristol. Cd.8056. 1915

Second Interim Report on an Investigation of Industrial Fatigue by Physiological Methods, by A. F. Stanley Kent, M.A., D.Sc. (Oxon), Henry Overton Wills Professor of Physiology in the University of Bristol. Cd. 8335. 1916

PA Hansard's Parliamentary Debates
HC Deb House of Commons Debates
HL Deb House of Lords debates

Journals

The Boston Medical and Surgical Journal
British Medical Journal
The Journal of Balneology and Climatology
Journal of the American Medical Association
The Lancet
Railway Magazine

Newspapers and magazines

ProQuest Historical Newspapers (PQHN) https://about.proquest.com/products-services/pq-hist-news.html

London Society, August 1898
The Manchester Guardian, 11 July 1896
The Observer, 20 August 1871; 1 July 1877
The Quiver, June 1915
William H Harper's New Monthly Magazine 63 (1881)

The British Newspaper Archive (BNA) http://www.britishnewspaperarchive.co.uk

The Bath Chronicle, 19 August 1897
The Bolton Chronicle, 16 September 1871
The Bristol Mercury, 27 December 1895
The Cambridge Independent Press, 15 August 1902; 3 July 1903; 27 August 1909
Cheltenham Chronicle and Graphic, 28 June 1902
The Daily News, 30 May 1871
The Derby Mercury, 10 January 1872; 5 June 1878
The Falkirk Herald and Linlithgow Journal, 9 May 1878
The Hull Packet and East Riding Times, 25 June 1880
The Manchester Courier, 26 August 1899
The Morning Chronicle, 30 September 1857
The North-Eastern Daily Gazette, 20 April 1889; 10 June 1889
The Penny Illustrated Paper, 6 August 1870
Saturday Review of Politics, Literature, Science and Art, 30 December 1899
The Star Guernsey, 8 September 1888
The Western Times, 28 May 1901; 5 April 1904
The Whitby Gazette, 2 August 1907

Secondary sources

Adams, Jane M. *Healing with Water: English Spas and the Water Cure, 1840–1960.* Manchester: Manchester University Press, 2015.

Ahmed, Sara. 'Affective Economies'. *Social Text* 22, no. 2 (2004): 117–39.

Alberti, Fay Bound. 'Introduction: Medical History and Emotion Theory'. In Fay Bound Alberti (ed.), *Medicine, Emotion and Disease, 1700–1950*, pp. xiii–xxviii. New York: Springer, 2006.

Alberti, Fay Bound (ed.). *Medicine, Emotion and Disease, 1700–1950.* New York: Springer, 2006.

Alberti, Fay Bound. 'Bodies, Hearts, and Minds: Why Emotions Matter to Historians of Science and Medicine'. *Isis* 100, no. 4 (2009): 798–810.

Appadurai, Arjun (ed.). *The Social Life of Things.* Cambridge: Cambridge University Press, 1986.

Bailey, Peter. *Leisure and Class in Victorian England: Rational Recreation and the Contest for Control, 1830–1885.* Buffalo: Routledge & Kegan Paul Books, 1978.

Bailey, Peter. 'Breaking the Sound Barrier: A Historian Listens to Noise'. *Body & Society* 2, no. 2 (1996): 49–66.

Baker, Amy. '7 Glaring Signs You Need a Holiday', *Metro*, 16 June 2017, http://metro.co.uk/2017/06/16/7-glaring-signs-you-need-a-holiday-6687106/(last accessed 20 February 2019).

Baker, William J. 'The Leisure Revolution in Victorian England: A Review of Recent Literature'. *Journal of Sport History* 6, no. 3 (1979): 76–87.

Bakhtin, Mikhail. *Rabelais and his World.* Translated by Helene Iswolsky. Bloomington: Indiana University Press, 1984.

Barbalet, Jack M. 'Boredom and Social Meaning'. *The British Journal of Sociology* 50, no. 4 (1999): 631–46.

Barton, Susan. 'Why We're All Going on a Summer Holiday: The Role of the Working-Class Organisations in the Development of Popular Tourism, 1850–1950'. PhD diss., De Montfort University, 1999.

Barton, Susan. *Working-Class Organisations and Popular Tourism, 1840–1970.* Manchester: Manchester University Press, 2005.

Barton, Susan. *Healthy Living in the Alps: Sanatoria and the Origins of Winter Sports Tourism in Switzerland, 1860–1914.* Manchester: Manchester University Press, 2009.

Baudelaire, Charles. *Paris Spleen.* Translated by Louise Varese. New York: New Directions Books, 1970 [1869].

Baudrillard, Jean. *The Consumer Society: Myths and Structures.* London: Sage, 1998.

Beaven, Brad. *Leisure, Citizenship and Working-Class Men in Britain, 1850–1945.* Manchester: Manchester University Press, 2005.

Beckerson, John, and John K. Walton. 'Selling Air: Marketing the Intangible at British Resorts'. In John K. Walton (ed.), *Histories of Tourism: Representation, Identity and Conflict*, pp. 55–68. Clevedon: Cromwell Press, 2005.

Beetham, Margaret. '"Oh! I Do Like to Be Beside the Seaside!': Lancashire Seaside Publications'. *Victorian Periodicals Review* 42, no. 1 (2009): 24–36.

Bell, Claudia, and John Lyall. *The Accelerated Sublime: Landscape, Tourism and Identity.* Westport: Praeger Publishers, 2002.

Benger Alaluf, Yaara. '"It Is All Included—Without the Stress": Exploring the Production of Relaxation in Club Med Seaside Resorts'. In Eva Illouz (ed.), *Emotions as Commodities: Capitalism, Consumption and Authenticity*, pp. 33–55. London: Routledge, 2017.

Benger Alaluf, Yaara, and Eva Illouz. 'Emotions in Consumer Studies'. In Frederick F. Wherry and Ian Woodward (eds), *The Oxford Handbook of Consumption*, pp. 239–52. Oxford: Oxford University Press, 2019.

Berger, Arthur Asa. *The Objects of Affection: Semiotics and Consumer Culture*. New York: Palgrave Macmillan, 2010.

Berghoff, Hartmut. 'From Privilege to Commodity? Modern Tourism and the Rise of the Consumer Society'. In Hartmut Berghoff, Barbara Korte, Ralf Schneider, and Christopher Harvie (eds), *The Making of Modern Tourism: The Cultural History of the British Experience, 1600–2000*, pp. 159–80. London: Palgrave Macmillan, 2002.

Berghoff, Hartmut, Barbara Korte, Ralf Schneider, and Christopher Harvie (eds). *The Making of Modern Tourism: The Cultural History of the British Experience, 1600–2000*. London: Palgrave Macmillan, 2002.

Berrios, German E. *The History of Mental Symptoms: Descriptive Psychopathology Since the Nineteenth Century*. Cambridge: Cambridge University Press, 1996.

Berryman, Bryan. *Scarborough as It Was*. Nelson: Hendon Publishing Company Limited, 1971.

Bienefeld, Manfred A. *Working Hours in British Industry: An Economic History*. London: Weidenfeld and Nicolson [for the] London School of Economics, 1972.

Boddice, Rob. *The Science of Sympathy: Morality, Evolution, and Victorian Civilization*. Urbana: University of Illinois Press, 2016.

Boddice, Rob. 'The History of Emotions: Past, Present, Future'. *Revista de Estudios Sociales* 62 (2017): 10–15.

Boorstin, Daniel. 'From Traveler to Tourist: The Lost Art of Travel'. In *The Image: Or What Happened to the American Dream*, pp. 77–117. New York: Vintage Books, 1992 [1961].

Borsay, Peter. 'The Emergence of a Leisure Town: Or an Urban Renaissance?' *Past & Present* 126 (1990): 189–96.

Borsay, Peter. *A History of Leisure: The British Experience Since 1500*. Hampshire: Palgrave Macmillan, 2006.

Borsay, Peter. 'From Port to Resort: Tenby and Narratives of Transition, 1760–1914'. In Peter Borsay and John K. Walton (eds), *Resorts and Ports: European Seaside Towns since 1700*, pp. 86–112. Bristol: Channel View Publications, 2011.

Borsay, Peter. 'Town or Country? British Spas and the Urban–Rural Interface'. *Journal of Tourism History* 4, no. 2 (2012): 155–69.

Borsay, Peter. 'A Room with a View: Visualising the Seaside, c. 1750–1914'. *Transactions of the Royal Historical Society* 23 (2013): 175–201.

Borsay, Peter, and John K. Walton. 'Introduction: The Resort–Port Relationship'. In Peter Borsay and John K. Walton (eds), *Resorts and Ports: European Seaside Towns since 1700*, pp. 1–17. Bristol: Channel View Publications, 2011.

Borsay, Peter, and John K. Walton (eds). *Resorts and Ports: European Seaside Towns Since 1700*. Bristol: Channel View Publications, 2011.

Bourdieu, Pierre. *Outline of a Theory of Practice*. Cambridge: Cambridge University Press, 1977.

Bourke, Joanna. *Fear: A Cultural History*. London: Virago Press, 2005.

Bourke, Joanna. *The Story of Pain: From Prayer to Painkillers*. Oxford: Oxford University Press, 2014.

Bridge, Adrian. 'Holidays Are Good for You—It's Official'. *The Telegraph*, 31 January 2013, accessed 20 February 2019. http://www.telegraph.co.uk/travel/news/Holidays-are-good-for-you-its-official/.

Brodie, Allan. 'Towns of "Health and Mirth": The First Seaside Resorts, 1730–1769'. In Peter Borsay and John K. Walton (eds), *Resorts and Ports: European Seaside Towns Since 1700*, pp. 18–32. Bristol: Channel View Publications, 2011.

Buck-Morss, Susan. *The Dialectics of Seeing: Walter Benjamin and the Arcades Project*. Cambridge: MIT Press, 1991.

Buettner, Linda, Mona Shattell, and Madeleine Reber. 'Working Hard to Relax: Improving Engagement in Leisure Time Activities for a Healthier Work–Life Balance'. *Issues in Mental Health Nursing* 32, no. 4 (2011): 269–70.

Burkitt, Ian. 'Social Relationships and Emotions'. *Sociology* 31, no. 1 (1997): 37–55.

Cabanas, Edgar. 'Experiencing Designs and Designing Experiences: Emotions and Theme Parks from a Symbolic Interactionist Perspective'. *Journal of Destination Marketing & Management* 16 (2020).

Cabanas, Edgar. 'Rekindling Individualism, Consuming Emotions: Constructing "Psytizens" in the Age of Happiness'. *Culture & Psychology* 22, no. 3 (2016): 467–80.

Caldwell, Linda L. 'Leisure and Health: Why Is Leisure Therapeutic?' *British Journal of Guidance & Counselling* 33, no. 1 (2005): 7–26.

Campbell, Colin. 'Conspicuous Confusion? A Critique of Veblen's Theory of Conspicuous Consumption'. *Sociological Theory* 13 (1995): 37–47.

Campbell, Colin. *The Romantic Ethic and the Spirit of Modern Consumerism*. York: Alcuin Academics, 2005.

Cantor, David. 'The Contradictions of Specialization: Rheumatism and the Decline of the Spa in Inter-War Britain'. *Medical History. Supplement*, 34, no. 10 (1990): 127–44.

Cater, Carl, and Paul Cloke. 'Bodies in Action: The Performativity of Adventure Tourism'. *Anthropology Today* 23, no. 6 (2007): 13–16.

Chandola, Tripta. 'Listening into Others: Moralising the Soundscapes in Delhi'. *International Development Planning Review* 34, no. 4 (2012): 391–408.

Chen, Chun-Chu, Wei-Jue Huang, and James F. Petrick. 'Holiday Recovery Experiences, Tourism Satisfaction and Life Satisfaction—Is There a Relationship?' *Tourism Management* 53 (2016): 140–7.

Cohen, Erik. 'A Phenomenology of Tourist Experiences'. *Sociology* 13, no. 2 (1979): 179–201.

Cohen, Erik. *Contemporary Tourism: Diversity and Change*. Amsterdam: Elsevier, 2004.

Cohen, Scott A., Girish Prayag, and Miguel Moital. 'Consumer Behaviour in Tourism: Concepts, Influences and Opportunities'. *Current Issues in Tourism* 17, no. 10 (2014): 872–909.

Colema, Naomi. 'Why Holidays are Healthy'. *Daily Mail*, last accessed 20 February 2019. http://www.dailymail.co.uk/health/article-60094/Why-holidays-healthy.html.

Coleman, Simon, and Mike Crang (eds). *Tourism: Between Place and Performance*. New York: Berghahn Books, 2002.

Coley, Noel G. 'Physicians and the Chemical Analysis of Mineral Waters in Eighteenth-Century England'. *Medical History* 26, no. 2 (1982): 123–44.

Coombs, Hugh M., and John Richard Edwards. 'The Financial Reporting Practices of British Municipal Corporations 1835–1933: A Study in Accounting Innovation'. *Accounting and Business Research* 25, no. 98 (1995): 93–105.

Corbin, Alain. *The Lure of the Sea: The Discovery of the Seaside in the Western World, 1750–1840*. Translated by Jocelyn Phelps. Berkeley: University of California Press, 1994.
Cossic, Annick. 'Introduction'. In Patrick Galliou and Annick Cossic (eds), *Spas in Britain and in France in the Eighteenth and Nineteenth Centuries*, pp. xiii–xx. Newcastle: Cambridge Scholars Press, 2006.
Cross, Gary S. 'The Quest for Leisure: Reassessing the Eight-Hour Day in France'. *Journal of Social History* 18, no. 2 (1984): 195–216.
Cross, Gary S. *A Quest for Time: The Reduction of Work in Britain and France, 1840–1940*. Berkeley: University of California Press, 1989.
Cross, Gary S. 'Time, Money, and Labor History's Encounter with Consumer Culture'. *International Labor and Working-Class History* 43 (1993): 2–17.
Cross, Gary S. 'A Right to Be Lazy? Busyness in Retrospective'. *Social Research* 72, no. 2 (2005): 263–86.
Cross, Gary S., and John K. Walton. *The Playful Crowd: Pleasure Places in the Twentieth Century*. New York: Columbia University Press, 2005.
Crossick, Geoffrey. 'The Emergence of the Lower Middle Class in Britain: A Discussion'. In Geoffrey Crossick (ed.), *The Lower Middle Class in Britain, 1870–1914*, pp. 11–60. New York: St Martin's Press, 1977.
Dahlberg, Lincoln. 'The Habermasian Public Sphere: Taking Difference Seriously?' *Theory and Society* 34, no. 2 (2005): 111–36.
Danziger, Kurt. *Naming the Mind: How Psychology Found Its Language*. London: Sage, 1997.
Daston, Lorraine. 'The Moral Economy of Science'. *Osiris* 10, no. 1 (1995): 2–24.
Daston, Lorraine. 'Attention and the Values of Nature in the Enlightenment'. In Lorraine Daston and Fernando Vidal (eds), *The Moral Authority of Nature*, pp. 100–26. Chicago: University of Chicago Press, 2010.
D'Cruze, Shani. 'Approaching the History of Rape and Sexual Violence: Notes towards Research'. *Women's History Review* 1, no. 3 (1992): 377–97.
De Bloom, Jessica, Sabine A. E. Geurts, and Michiel A. J. Kompier. 'Vacation (After-) Effects on Employee Health and Well-Being, and the Role of Vacation Activities, Experiences and Sleep'. *Journal of Happiness Studies* 14, no. 2 (2013): 613–33.
Dixon, Thomas. 'The Psychology of the Emotions in Britain and America in the Nineteenth Century: The Role of Religious and Antireligious Commitments'. *Osiris 16* (2001): 288–320.
Dixon, Thomas. *From Passions to Emotions: The Creation of a Secular Psychological Category*. Cambridge: Cambridge University Press, 2003.
Dixon, Thomas. 'Patients and Passions: Languages of Medicine and Emotion, 1789–1850'. In Fay Bound Alberti (ed.), *Medicine, Emotion and Disease, 1700–1950*, pp. 22–52. New York: Springer, 2006.
Donauer, Sabine. 'Emotions at Work—Working on Emotions: The Production of Economic Selves in Twentieth-Century Germany'. PhD diss., Free University of Berlin, 2013.
Durie, Alastair. 'Medicine, Health and Economic Development: Promoting Spa and Seaside Resorts in Scotland c. 1750–1830'. *Medical History* 47, no. 2 (2003): 195–216.
Durie, Alastair. *Water Is Best: The Hydros and Health Tourism in Scotland, 1840–1940*. Edinburgh: John Donald, 2006.

Durie, Alastair. 'Tourism and the Railways in Scotland: The Victorian and Edwardian Experience'. In Anne K. B. Evans and John V. Gough (eds), *The Impact of the Railway on Society in Britain: Essays in Honour of Jack Simmons*, pp. 199–209. London: Routledge, 2016.

Durie, Alastair. *Scotland and Tourism: The Long View, 1700–2015*. London: Routledge, 2017.

Edensor, Tim. 'Performing Tourism, Staging Tourism: (Re)producing Tourist Space and Practice'. *Tourist Studies* 1, no. 1 (2001): 59–81.

Ehrenreich, Barbara, and Deirdre English. *For Her Own Good: Two Centuries of the Experts' Advice to Women*. New York: Anchor Books, 1978.

Eitler Pascal, Stephanie Olsen, and Uffa Jensen. 'Introduction'. In Ute Frevert, Pascal Eitler, Stephanie Olsen, Uffa Jensen, Margrit Pernau, Daniel Brückenhaus, Magdalena Beljan, Benno Gammerl, Anja Laukötter, Bettina Hitzer, Jan Plamper, Juliane Brauer, and Joachim C. Häberlen (eds), *Learning How to Feel: Children's Literature and Emotional Socialization, 1870–1970*, pp. 1–20. Oxford: Oxford University Press, 2014.

Farrant, Sue. 'London by the Sea: Resort Development on the South Coast of England 1880–1939'. *Journal of Contemporary History* 22, no. 1 (1987): 137–62.

Fassin, Didier. 'Compassion and Repression: The Moral Economy of Immigration Policies in France'. *Cultural Anthropology* 20, no. 3 (2005): 362–87.

Fassin, Didier. 'Les économies morales revisitées'. *Annales. Histoire, Sciences Sociales* 64, no. 6 (2009): 1237–66.

Featherstone, Mike. *Consumer Culture and Postmodernism*. London: Sage, 1991.

Featherstone, Mike. 'The Body in Consumer Culture'. In Jessica R. Johnston (ed.), *The American Body in Context: An Anthology*, pp. 79–102. Wilmington: Scholarly Resources Inc., 2001.

Foley, Ronan. *Healing Waters: Therapeutic Landscapes in Historic and Contemporary Ireland*. Farnham: Ashgate Publishing, 2010.

Franklin, Adrian, and Mike Crang. 'The Trouble with Tourism and Travel Theory?'. *Tourist Studies* 1 no. 1 (2001): 5–22.

Freudenberger, Herman, and Gaylord Cummins. 'Health, Work, and Leisure Before the Industrial Revolution'. *Explorations in Economic History* 13, no. 1 (1976): 1–12.

Frevert, Ute. *Emotions in History: Lost and Found*. Budapest: Central European University Press, 2011.

Frevert, Ute. 'Passions, Preferences, and Animal Spirits: How Does Homo Oeconomicus Cope with Emotions?' In Frank Biess and Daniel M. Gross (eds), *Science and Emotions after 1945: A Transatlantic Perspective*, pp. 300–17. Chicago: University of Chicago Press, 2014.

Frevert, Ute, Monique Scheer, Pascal Eitler, Anne Schmidt, Bettina Hitzer, Nina Verheyen, Benno Gammerl, Christian Bailey, and Margrit Pernau. *Emotional Lexicons: Continuity and Change in the Vocabulary of Feeling 1700–2000*. Oxford: Oxford University Press, 2014.

Friedberg, Anne. *Window Shopping: Cinema and the Postmodern*. Berkeley: University of California Press, 1993.

Frisby, David. *Fragments of Modernity: Theories of Modernity in the Work of Simmel, Kracauer and Benjamin*. Plymouth: Routledge, 2013.

Fritz, Charlotte, and Sabine Sonnentag. 'Recovery, Well-Being, and Performance-Related Outcomes: The Role of Workload and Vacation Experiences'. *Journal of Applied Psychology* 91, no. 4 (2006): 936–45.

Gammerl, Benno. 'Transitory Feelings? On Challenges and Trends Within the History of Emotions'. *Contemporanea* 17, no. 2 (2014): 335–44.

Gibson, Chris, and Lily Kong. 'Cultural Economy: A Critical Review'. *Progress in Human Geography* 29, no. 5 (2005): 541–61.

Gijswijt-Hofstra, Marijke. 'Introduction: Cultures of Neurasthenia from Beard to the First World War'. In Marijke Gijswijt-Hofstra and Roy Porter (eds), *Cultures of Neurasthenia from Beard to the First World War*, pp. 1–31. Amsterdam: Rodopi, 2001.

Gijswijt-Hofstra, Marijke, and Roy Porter (eds). *Cultures of Neurasthenia from Beard to the First World War*. Amsterdam: Rodopi, 2001.

Goering, Laura. '"Russian Nervousness": Neurasthenia and National Identity in Nineteenth-Century Russia'. *Medical History* 47, no. 1 (2003): 23–46.

Goldstein, Jan E. *Console and Classify: The French Psychiatric Profession in the Nineteenth Century*. Cambridge: Cambridge University Press, 1987.

Goodman, Douglas J., and Mirelle Cohen. 'From Consumption to Consumer Culture'. In Douglas J. Goodman and Mirelle Cohen (eds), *Consumer Culture: A Reference Handbook*, pp. 1–26. Santa Barbara: ABC-CLIO, 2004.

Gosling, Francis G., and Joyce M. Ray. 'The Right to Be Sick: American Physicians and Nervous Patients, 1885–1910'. *Journal of Social History* 20, no. 2 (1986): 251–67.

Götz, Norbert. '"Moral Economy": Its Conceptual History and Analytical Prospects'. *Journal of Global Ethics* 11, no. 2 (2015): 147–62.

Gray, Fred. *Designing the Seaside: Architecture, Society and Nature*. London: Reaktion Books, 2006.

Gunning, Tom. 'An Aesthetic of Astonishment: Early Film and the (In)Credulous Spectator'. In Philip Simpson, Andrew Utterson, and Karen J. Shepherdson (eds), *Film Theory: Critical Concepts in Media and Cultural Studies*, vol. 3, pp. 78–95. London: Routledge, 2004.

Habermas, Jürgen. *The Structural Transformation of the Public Sphere: An Inquiry into a Category of Bourgeois Society*. Translated by Thomas Burger. Cambridge: MIT Press, 1991.

Hamlin, Christopher. 'Chemistry, Medicine, and the Legitimization of English Spas, 1740–1840'. *Medical History Supplement* 10 (1990): 67–81.

Harwood, Stephen, and Dahlia El-Manstrly. 'The Performativity Turn in Tourism'. *University of Edinburgh Business School Working Paper Series* 12, no. 5 (2012): 1–18.

Heeley, John. 'Planning for Tourism in Britain: An Historical Perspective'. *Town Planning Review* 52, no. 1 (1981): 61–79.

Hembry, Phyllis May. *The English Spa 1560–1815: A Social History*. London: The Athlone Press, 1990.

Hembry, Phyllis May. *British Spas from 1815 to the Present: A Social History*. New Jersey: Fairleigh Dickinson University Press, 1997.

Hennock, E. P. 'IV. Finance and Politics in Urban Local Government in England, 1835–1900'. *The Historical Journal* 6, no. 2 (1963): 212–25.

Hern, Anthony. *The Seaside Holiday: The History of the English Seaside Resort*. London: The Cresset Press, 1967.

Hindley, Diana, and Geoffrey Hindley. *Advertising in Victorian England, 1837–1901*. London: Hodder Wayland, 1972.

Hochschild, Arlie Russell. *The Managed Heart: Commercialization of Human Feeling*. Berkeley: University of California Press, 1983.

'Holiday & Vacation Countdown Timer. Developer: Pro Sellers World LLC'. *iTunes*, 7 March 2017, https://itunes.apple.com/de/app/holiday-vacation-countdown-timer-event-widget/id1034678674?l=en&mt=8 (last accessed 20 February 2019).

'Holiday Countdown. Developer: Holiday Extras Ltd'. *GooglePlay*, 16 January 2017, https://itunes.apple.com/de/app/holiday-vacation-countdown-timer-event-widget/id1034678674?l=en&mt=8 (last accessed 20 February 2019)

Hosgood, Christopher P. '"Mercantile Monasteries": Shops, Shop Assistants, and Shop Life in Late-Victorian and Edwardian Britain'. *Journal of British Studies* 38, no. 3 (1999): 322–52.

Houghton, Walter E. *The Victorian Frame of Mind, 1830–1870*. New Haven: Yale University Press, 1957.

Huberman, Michael. 'Working Hours of the World Unite? New International Evidence of Worktime, 1870–1913'. *The Journal of Economic History* 64, no. 4 (2004): 964–1001.

Huggins, Mike. 'More Sinful Pleasures? Leisure, Respectability and the Male Middle Classes in Victorian England'. *Journal of Social History* 33, no. 3 (2000): 585–600.

Illouz, Eva. *Consuming the Romantic Utopia: Love and the Cultural Contradictions of Capitalism*. Berkeley, University of California Press, 1997.

Illouz, Eva. *Cold Intimacies: The Making of Emotional Capitalism*. Cambridge: Polity Press, 2007.

Illouz, Eva. *Saving the Modern Soul: Therapy, Emotions, and the Culture of Self-Help*. Berkeley: University of California Press, 2008.

Illouz, Eva. 'Emotions, Imagination and Consumption: A New Research Agenda'. *Journal of Consumer Culture* 9, no. 3 (2009): 377–413.

Illouz, Eva (ed.). *Emotions as Commodities: Capitalism, Consumption and Authenticity*. London: Routledge, 2017.

Illouz, Eva. 'Introduction: Emodities or the Making of Emotional Commodities'. In Eva Illouz (ed.), *Emotions as Commodities: Capitalism, Consumption and Authenticity*, pp. 1–29. London: Routledge, 2017.

Illouz, Eva. 'Toward a Post-normative Critique of Emotional Authenticity'. In Eva Illouz (ed.), *Emotions as Commodities: Capitalism, Consumption and Authenticity*, 197–213. London: Routledge, 2017.

Illouz, Eva, and Yaara Benger Alaluf. 'Emotions and Consumption'. In Dan Cook and J. Michael Ryan (eds), *The Wiley-Blackwell Encyclopedia of Consumption and Consumer Studies*, 1st edn, pp. 263–8. Oxford: John Wiley & Sons, 2015.

Illouz, Eva, and Yaara Benger Alaluf. 'Le capitalisme émotionnel'. In Alain Corbin, Jean-Jacques Courtine, and Georges Vigarello (eds), *Histoire des émotions vol. 3: De la fin du XIXe siècle à nos jours (1890–2013)*, pp. 56–75. Paris: Seuil, 2017.

Ilmonen, Kaj. 'Want, Need and Commodity'. In Kaj Ilmonen, Pekka Sulkunen, Keijo Rahkonen, Jukka Gronow, Arto Noro, and Alan Warde (eds), *A Social and Economic Theory of Consumption*, pp. 45–72. London: Palgrave Macmillan, 2010.

Inglis, Fred. *The Delicious History of the Holiday*. London: Routledge, 2005.

Iyengar, Sheena S., and Mark R. Lepper. 'When Choice Is Demotivating: Can One Desire Too Much of a Good Thing?' *Journal of Personality and Social Psychology* 79, no. 6 (2000): 995–1006.

Jackson, Mark (ed.). *Stress in Post-War Britain, 1945–85*. Oxon: Routledge, 2015.

Jacyna, Leon Stephen. 'The Physiology of Mind, the Unity of Nature, and the Moral Order in Victorian Thought'. *The British Society for the History of Science* 14, no. 2 (1981): 109–32.

Jankovic, Vladimir. 'The Last Resort: A British Perspective on the Medical South, 1815–1870'. *Journal of Intercultural Studies* 27, no. 3 (2006): 271–98.

Jenkins, Henry. *Convergence Culture: Where Old and New Media Collide*. New York: New York University Press, 2006.

Johannison, Karin. 'Modern Fatigue: A Historical Perspective'. In Bengt B. Arnetz and Rolf Ekman (eds), *Stress in Health and Disease*, pp. 3–19. Weinheim: Wiley-VCH Verlag, 2006.

Johnson, Mary Orr. 'The Insane in 19th Century Britain: A Statistical Analysis of a Scottish Insane Asylum'. *Historical Social Research* 17, no. 3 (1992): 3–20.

Jones, Linda J. 'Public Pursuit of Private Profit? Liberal Businessmen and Municipal Politics in Birmingham, 1865–1900'. *Business History* 25, no. 3 (1983): 240–59.

Kane, Josephine. 'Edwardian Amusement Parks: The Pleasure Garden Reborn?'. In Jonathan Conlin (ed.), *The Pleasure Garden, from Vauxhall to Coney Island*, pp. 217–46. Philadelphia: University of Pennsylvania Press, 2012.

Kane, Josephine. *The Architecture of Pleasure: British Amusement Parks 1900–1939*. London: Routledge, 2016.

Kaufmann, Doris. 'Neurasthenia in Wilhelmine Germany: Culture, Sexuality, and the Demands of Nature'. In Marijke Gijswijt-Hofstra and Roy Porter (eds), *Cultures of Neurasthenia from Beard to the First World War*, pp. 161–76. Amsterdam: Rodopi, 2001.

Koenker, Diane. *Club Red: Vacation Travel and the Soviet Dream*. Ithaca: Cornell University Press, 2013.

Leshem, Dotan. 'Oikonomia Redefined'. *Journal of the History of Economic Thought* 35, no. 1 (2013): 43–61.

Leshem, Dotan. *The Origins of Neoliberalism: Modeling the Economy from Jesus to Foucault*. New York: Columbia University Press, 2016.

Leunig, Timothy. 'Time Is Money: A Re-assessment of the Passenger Social Savings from Victorian British Railways'. *The Journal of Economic History* 66, no. 3 (2006): 635–73.

Leys, Ruth. 'The Turn to Affect: A Critique'. *Critical Inquiry* 37, no. 3 (2011): 434–72.

Löfgren, Orvar. *On Holiday: A History of Vacationing*. Berkeley: University of California Press, 1999.

Loughran, Tracey. 'Hysteria and Neurasthenia in Pre-1914 British Medical Discourse and in Histories of Shell-Shock'. *History of Psychiatry* 19, no. 1 (2008): 25–46.

Lowerson, John, and John Myerscough. *Time to Spare in Victorian England*. Hassocks: Harvester Press, 1977.

Lowry, Candace. 'How Badly Do You Need A Vacation?'. *BuzzFeed*, 21 August 2014, last accessed 20 February 2019. https://www.buzzfeed.com/candacelowry/how-badly-do-you-need-a-vacation?utm_term=.impPZdMXX#.xfEKQAlWW.

Lynch, Alison. '17 Signs You Really, Really Need a Holiday', *Metro*, 23 August 2014, last accessed 20 February 2019. http://metro.co.uk/2014/08/23/16-signs-you-really-really-need-a-holiday-4842512/.

MacCannell, Dean. 'Staged Authenticity: Arrangements of Social Space in Tourist Settings', *American Journal of Sociology* 79, no. 3 (1973): 589–603.

MacCannell, Dean. *The Tourist: A New Theory of the Leisure Class*. Berkeley: University of California Press, 1976.

Magnusson, Warren. 'Bourgeois Theories of Local Government'. *Political Studies* 34, no. 1 (1986): 1–18.

Maines, Rachel P. *The Technology of Orgasm: 'Hysteria', the Vibrator, and Women's Sexual Satisfaction*. Baltimore: The Johns Hopkins University Press, 1999.

Marcuse, Herbert. *One-Dimensional Man: Studies in the Ideology of Advanced Industrial Society*. Oxon: Routledge, 2013 [1964].

Masters, Maria. '10 Signs You Really Need a Vacation'. *Health*, 31 July 2015, last accessed 20 February 2019. http://www.health.com/health/gallery/0,,20936703,00.html#when-to-take-a-day-off-0.

Matt, Susan. *Keeping Up with the Joneses: Envy in American Consumer Society, 1890–1930.* Philadelphia: University of Pennsylvania Press, 2003.

Matt, Susan. 'Current Emotion Research in History: Or, Doing History from the Inside Out'. *Emotion Review* 3, no. 1 (2011): 117–24.

McClelland, Keith. 'Time to Work, Time to Live: Some Aspects of Work and the Reformation of Class in Britain, 1850–1880'. In Patrick Joyce (ed.), *The Historical Meanings of Work*, pp. 180–209. Cambridge: Cambridge University Press, 1987.

McInnes, Angus. 'The Emergence of a Leisure Town: Shrewsbury 1660–1760'. *Past & Present* 120 (1988): 53–87.

McIvor, Arthur J. 'Employers, the Government, and Industrial Fatigue in Britain, 1890–1918'. *Occupational and Environmental Medicine* 44, no. 11 (1987): 724–32.

Meller, Helen. *Leisure and the Changing City 1870–1914.* Oxon: Routledge, 2013.

Metcalfe, Alan. *Leisure and Recreation in a Victorian Mining Community: The Social Economy of Leisure in North-East England, 1820–1914.* London: Taylor & Francis, 2005.

Millward, Roy. 'Railways and the Evolution of Welsh Holiday Resorts'. In Anne K. B. Evans and John V. Gough (eds), *The Impact of the Railway on Society in Britain: Essays in Honour of Jack Simmons*, pp. 211–23. London: Routledge, 2016.

Morat, Daniel. 'Urban Soundscape and Acoustic Innervation around 1900'. In Robert Beck, Ulrike Krampl, and Emmanuelle Reaillaud (eds), *Les cinq sens de la ville. Du moyen âge à nos jours*, pp. 71–83. Paris: Presses Universitaires François Rabelais, 2013.

Morat, Daniel (ed.). *Sounds of Modern History: Auditory Cultures in 19th-and 20th-Century Europe.* New York: Berghahn Books, 2014.

Mueller, Hansruedi, and Eveline Lanz Kaufmann. 'Wellness Tourism: Market Analysis of a Special Health Tourism Segment and Implications for the Hotel Industry'. *Journal of Vacation Marketing* 7, no. 1 (2001): 5–17.

Munro, Ian. *The Figure of the Crowd in Early Modern London: The City and Its Double.* New York: Palgrave Macmillan, 2005.

Negri, Antonio, and Michael Hardt. 'Value and Affect'. *boundary 2* 26, no. 2 (1999): 77–88.

Neve, Michael. 'Public Views of Neurasthenia: Britain, 1880–1930'. In Marijke Gijswijt-Hofstra and Roy Porter (eds), *Cultures of Neurasthenia from Beard to the First World War*, pp. 141–60. Amsterdam: Rodopi, 2001.

Newmeyer, Trent S. '"Moral Renovation and Intellectual Exaltation": Thomas Cook's Tourism as Practical Education'. *Journal of Tourism and Cultural Change* 6, no. 1 (2008): 1–16.

O'Connell, Brid. 'Quiz: How Much Do You Need a Holiday?' *Thomas Cook Airlines*, 10 February 2017, last accessed 20 February 2019. https://blog.thomascookairlines.com/quiz-much-need-holiday/.

Palmowsk, Jan. 'Liberalism and Local Government in Late Nineteenth-Century Germany and England'. *The Historical Journal* 45, no. 2 (2002): 381–409.

Palomera, Jaime, and Theodora Vetta. 'Moral Economy: Rethinking a Radical Concept'. *Anthropological Theory* 16, no. 4 (2016): 413–32.

Parkins, Wendy. 'Trust Your Senses? An Introduction to the Victorian Sensorium'. *Australasian Journal of Victorian Studies* 14, no. 2 (2010): 1–7.

Perkin, Harold J. 'The "Social Tone" of Victorian Seaside Resorts in the North-West'. *Northern History* 11, no. 1 (1976): 180–94.

Pernau, Margrit. 'Male Anger and Female Malice: Emotions in Indo-Muslim Advice Literature'. *History Compass* 10, no. 2 (2012): 119–28.

Pernau, Margrit. 'Provincializing Concepts: The Language of Transnational History'. *Comparative Studies of South Asia, Africa and the Middle East* 36, no. 3 (2016): 483–99.

Pietikainen, Petteri. *Neurosis and Modernity: The Age of Nervousness in Sweden*. Leiden: Brill, 2007.

Pimlott, John Alfred Ralph. 'The English Seaside'. *English* 10, no. 59 (1955): 166–71.

Pimlott, John Alfred Ralph. *The Englishman's Holiday: A Social History*. Hassocks: Harvester Press, 1976 [1947].

Pine, Joseph, and James H. Gilmore. *Experience Economy: Work is Theatre & Every Business a Stage*. Boston: Harvard Business School Press, 1999.

Plamper, Jan. *The History of Emotions: An Introduction*. Oxford: Oxford University Press, 2015.

Poole, Robert. 'Oldham Wakes'. In John K. Walton and James Walvin (eds), *Leisure in Britain 1780–1939*, pp. 71–98. Manchester: Manchester University Press, 1983.

Poon, Angelia. *Enacting Englishness in the Victorian Period: Colonialism and the Politics of Performance*. Farnham: Ashgate Publishing, Ltd, 2008.

Porter, Roy. *Disease, Medicine and Society in England, 1550–1860*. Cambridge: Cambridge University Press, 1995.

Porter, Roy. 'Nervousness, Eighteenth and Nineteenth Century Style: From Luxury to Labour'. *Clio medica* 63 (2000): 31–49.

Presley, John Woodrow. 'Neurasthenia and the Cure of Literature: Robert Graves, Siegfried Sassoon, Andy Collins'. *JAC* 30, no. 1/2 (2010): 269–313.

Quan, Shuai, and Ning Wang. 'Towards a Structural Model of the Tourist Experience: An Illustration from Food Experiences in Tourism'. *Tourism Management* 25, no. 3 (2004): 297–305.

Rabinbach, Anson. *The Human Motor: Energy, Fatigue, and the Origins of Modernity*. Berkeley: University of California Press, 1992.

Radkau, Joachim. *Das Zeitalter der Nervosität: Deutschland zwischen Bismarck und Hitler*. München: Carl Hanser, 1998.

Ramirez, Rafael. 'Value Co-Production: Intellectual Origins and Implications for Practice and Research'. *Strategic Management Journal* 20, no. 1 (1999): 49–65.

Rapp, Dean. 'The Early Discovery of Freud by the British General Educated Public, 1912–1919'. *Social History of Medicine* 3, no. 2 (1990): 217–43.

Reddy, William M. 'Against Constructionism: The Historical Ethnography of Emotions'. *Current Anthropology* 38, no. 3 (1997): 327–51.

Redfoot, Donald L. 'Touristic Authenticity, Touristic Angst, and Modern Reality'. *Qualitative Sociology* 7, no. 4 (1984): 291–309.

Reid, Douglas A. 'The "Iron Roads" and "the Happiness of the Working Classes": The Early Development and Social Significance of the Railway Excursion'. *The Journal of Transport History* 17, no. 1 (1996): 57–73.

Reid, Douglas A. 'Weddings, Weekdays, Work and Leisure in Urban England 1791–1911: The Decline of Saint Monday Revisited'. *Past & Present* 153, no. 1 (1996): 135–63.

Richards, Thomas. *The Commodity Culture of Victorian England: Advertising and Spectacle, 1851–1914*. Stanford: Stanford University Press, 1991.

Roberts, Richard. 'The Corporation as Impresario: The Municipal Provision of Entertainment in Victorian and Edwardian Bournemouth'. In John K. Walton and James Walvin (eds), *Leisure in Britain 1780–1939*, pp. 136–57. Manchester: Manchester University Press, 1983.

Rojek, Chris. *Capitalism and Leisure Theory*. Oxon: Routledge, 2013.

Rose, Nikolas. *The Psychological Complex: Psychology, Politics and Society in England, 1869–1939*. London: Routledge & Kegan Paul, 1985.

Rudé, George F. E. *The Crowd in History: A Study of Popular Disturbances in France and England, 1730–848*. London: Serif, 2005.

Rylance, Rick. *Victorian Psychology and British Culture, 1850–1880*. Oxford: Oxford University Press, 2000.

Scheer, Monique. 'Are Emotions a Kind of Practice (and Is That What Makes Them Have a History)? A Bourdieuian Approach to Understanding Emotion'. *History and Theory* 51, no. 2 (2012): 193–220.

Schwartz, Barry. *The Paradox of Choice: Why More is Less*. New York: Ecco, 2004.

Schwarz, Ori. 'The Sound of Stigmatization: Sonic Habitus, Sonic Styles, and Boundary Work in an Urban Slum'. *American Journal of Sociology* 121, no. 1 (2015): 205–42.

Scott, James C. *The Moral Economy of the Peasant: Rebellion and Subsistence in Southeast Asia*. New Haven: Yale University Press, 1976.

Scull, Andrew (ed.). *Madhouses, Mad-Doctors, and Madmen: The Social History of Psychiatry in the Victorian Era*. Philadelphia: University of Pennsylvania Press, 1981.

Segrave, Kerry. *Beware the Masher: Sexual Harassment in American Public Places, 1880–1930*. Jefferson: McFarland, 2014.

Sengoopta, Chandak. 'A Mob of Incoherent Symptoms? Neurasthenia in British Medical Discourse, 1860–1920'. In Marijke Gijswijt-Hofstra and Roy Porter (eds), *Cultures of Neurasthenia from Beard to the First World War*, pp. 97–116. Amsterdam: Rodopi, 2001.

Shachak, Mattan. '(Ex)Changing Feelings: On the Commodification of Emotions in Psychotherapy'. In Eva Illouz (ed.), *Emotions as Commodities: Capitalism, Consumption and Authenticity*, pp. 147–72. London: Routledge, 2017.

Sharpley, Richard, and Philip R. Stone. 'Introduction: Experiencing Tourism, Experiencing Happiness?' In Richard Sharpley and Philip R. Stone (eds), *Contemporary Tourist Experience: Concepts and Consequences*, pp. 1–8. London: Routledge, 2012.

Shields, Rob. 'The "System of Pleasure": Liminality and the Carnivalesque at Brighton'. *Theory, Culture & Society* 7, no. 1 (1990): 39–72.

Shields, Rob. *Places on the Margin: Alternative Geographies of Modernity*. London: Routledge, 2013.

Shorter, Edward. *From Paralysis to Fatigue: A History of Psychosomatic Illness in the Modern Era*. New York: The Free Press, 2008.

Shorter, Edward. *How Everyone Became Depressed: The Rise and Fall of the Nervous Breakdown*. Oxford University Press, 2013.

Showalter, Elaine. *The Female Malady: Women, Madness, and Culture in England, 1830–1980*. New York: Pantheon Books, 1986.

Siméant, Johanna. 'Three Bodies of Moral Economy: The Diffusion of a Concept'. *Journal of Global Ethics* 11, no. 2 (2015): 163–75.

Simmons, Jack. 'Railways, Hotels, and Tourism in Great Britain 1839–1914'. *Journal of Contemporary History* 19, no. 2 (1984): 201–22.

Simmons, Jack. *The Victorian Railway*. London: Thames and Hudson, 1991.

Smith, Jacquelyn. '12 Signs You Desperately Need a Vacation from Work'. *Business Insider*, 2 January 2016, last accessed 20 February 2019. http://www.businessinsider.com/signs-you-need-a-vacation-2016-1?IR=T/#1-you-begin-to-look-like-your-passport-photo-1

Smith, Mark M. (ed.). *Hearing History: A Reader*. Athens: University of Georgia Press, 2004.

Smith, Melanie, and László Puczkó. *Health and Wellness Tourism* (Oxon: Routledge, 2009).

Soane, John. 'The Origin, Growth and Transformation of Maritime Resorts since 1840'. *Built Environment* 18 (1992): 12–26.

Soroka, Marina. *The Summer Capitals of Europe, 1814–1919*. Oxon: Routledge, 2017.

Spracklen, Karl. *The Meaning and Purpose of Leisure: Habermas and Leisure at the End of Modernity*. London: Palgrave Macmillan, 2009.
Stearns, Carol Z., and Peter N. Stearns.*Anger: The Struggle for Emotional Control in America's History*. Chicago: University of Chicago Press, 1989.
Stearns, Peter N. *Lives of Labor: Work in Maturing Industrial Society*. New York: Holmes and Meier Publishers, 1975.
Stearns, Peter N. *Consumerism in World History: The Global Transformation of Desire*. London: Routledge, 2006.
Stearns, Peter N., and Carol Z. Stearns. 'Emotionology: Clarifying the History of Emotions and Emotional Standards'. *The American Historical Review* 90, no. 4 (1985): 813–36.
Steward, Jill. 'The Affective Life of the Spa'. In David Picard and Mike Robinson (eds), *Emotion in Motion: Tourism, Affect and Transformation*, pp. 211–30. Farnham: Ashgate Publishing, 2012.
Steward, Jill. 'Moral Economies and Commercial Imperatives: Food, Diets and Spas in Central Europe: 1800–1914'. *Journal of Tourism History* 4, no. 2 (2012): 181–203.
Stobart, Jon, Andrew Hann, and Victoria Morgan. *Spaces of Consumption: Leisure and Shopping in the English Town, c. 1680–1830*. Oxon: Routledge, 2013.
Thompson, Edward P. 'Time, Work-Discipline, and Industrial Capitalism'. *Past and Present* 38, no. 1 (1967): 56–97.
Thompson, Edward P. 'The Moral Economy of the English Crowd in the Eighteenth Century'. *Past & Present* 50 (1971): 76–136.
Thomson, Mathew. 'Neurasthenia in Britain: An Overview'. In Marijke Gijswijt-Hofstra and Roy Porter (eds), *Cultures of Neurasthenia from Beard to the First World War*, pp. 77–95. Amsterdam: Rodopi, 2001.
Trentmann, Frank (ed.). *The Oxford Handbook of the History of Consumption*. Oxford: Oxford University Press, 2012.
Troschitz, Robert. 'Liminal Seaside? Working-Class Tourism in the Nineteenth Century'. In John Storey (ed.), *The Making of English Popular Culture*, pp. 104–17. Oxon: Routledge, 2016.
Turner, Louis, and John Ash. *The Golden Hordes: International Tourism and the Pleasure Periphery*. London: Constable & Robinson, 1975.
Urry, John. 'The "Consumption" of Tourism'. *Sociology* 24, no. 1 (1990): 23–35.
Urry, John. '*The Tourist Gaze* "Revisited"'. *American Behavioral Scientist* 36, no. 2 (1992): 172–86.
Urry, John. *Consuming Places*. London: Routledge, 1995.
Urry, John. *The Tourist Gaze*. 2nd edn. London: Sage Publications, 2002.
Urry, John. 'The Place of Emotions Within Place'. In Joyce Davidson, Liz Bondi and Mick Smith (eds), *Emotional Geographies*, pp. 77–83. Aldershot: Ashgate, 2007.
Voigt, Cornelia Graham Brown, and Gary Howat. 'Wellness Tourists: In Search of Transformation'. *Tourism Review* 66, no. 1/2 (2011): 16–30.
Voskuil, Lynn M. 'Feeling Public: Sensation Theater, Commodity Culture, and the Victorian Public Sphere'. *Victorian Studies* 44, no. 2 (2002): 245–74.
Vrettos, Athena. 'From Neurosis to Narrative: The Private Life of the Nerves in "Villette and Daniel Deronda"'. *Victorian Studies* 33, no. 4 (1990): 551–79.
Wainwright, David, and Michael Calnan. *Work Stress: The Making of a Modern Epidemic*. Buckingham: Open University Press, 2002.
Walkowitz, Judith R. 'Going Public: Shopping, Street Harassment, and Streetwalking in Late Victorian London'. *Representations* 62 (1998): 1–30.
Walton, John K. 'Residential Amenity, Respectable Morality, and the Rise of the Entertainment Industry: The Case of Blackpool'. *Literature and History* 1 (1975): 62–78.

Walton, John K. 'The Demand for Working-Class Seaside Holidays in Victorian England'. *The Economic History Review* 34, no. 2 (1981): 249–65.

Walton, John K. *The English Seaside Resort: A Social History 1750–1914*. Leicester: Leicester University Press, 1983.

Walton, John K. 'The Re-Making of a Popular Resort: Blackpool Tower and the Boom of the 1890s'. *The Local Historian* 24, no. 4 (1994): 194–205.

Walton, John K. 'Review of *The Lure of the Sea: The Discovery of the Seaside in the Western World 1750–1840*, by Alain Corbin'. *European History Quarterly* 25, no. 3 (1995): 450–52.

Walton, John K. 'Review of *British Spas from 1815 to the Present: A Social History*, by Phyllis Hembry'. *Urban History* 26, no. 2 (1999): 299–300.

Walton, John K. *The British Seaside: Holidays and Resorts in the Twentieth Century*. Manchester: Manchester University Press, 2000.

Walton, John K. 'British Tourism Between Industrialization and Globalization—An Overview'. In Hartmut Berghoff, Barbara Korte, Ralf Schneider, and Christopher Harvie (eds), *The Making of Modern Tourism: The Cultural History of the British Experience, 1600–2000*, pp. 109–32. London: Palgrave Macmillan, 2002.

Walton, John K. 'The History of British Spa Resorts: An Exceptional Case in Europe?'. *Transportes, Servicios y Telecomunicaciones* 20 (2011): 138–57.

Walton, John K. 'Health, Sociability, Politics and culture. Spas in History, Spas and History: An Overview'. *Journal of Tourism History*, 4, no. 1 (2012): 1–14.

Walton, John K., and James Walvin (eds). *Leisure in Britain 1780–1939*. Manchester: Manchester University Press, 1983.

Walvin, James. *Beside the Seaside: A Social History of the Popular Seaside Holiday*. London: Allen Lane, 1978.

Wang, Ning. *Tourism and Modernity: A Sociological Analysis*. Oxford: Pergamon, 2000.

Warde, Alan 'After Taste: Culture, Consumption and Theories of Practice'. *Journal of Consumer Culture* 14, no. 3 (2014): 279–303.

Watson, G. Llewellyn, and Joseph P. Kopachevsky. 'Interpretations of Tourism as Commodity'. *Annals of Tourism Research* 21, no. 3 (1994): 643–60.

Webb, Darren. 'Bakhtin at the Seaside: Utopia, Modernity and the Carnivalesque'. *Theory, Culture & Society* 22, no. 3 (2005): 121–38.

Weisz, George. 'Spas, Mineral Waters, and Hydrological Science in Twentieth-Century France'. *Isis* 92, no. 3 (2001): 451–83.

Weisz, George. 'Historical Reflections on Medical Travel'. *Anthropology & Medicine* 18, no. 1 (2011): 137–44.

Westman, Mina, and Dalia Etzion. 'The Impact of Vacation and Job Stress on Burnout and Absenteeism'. *Psychology & Health* 16, no. 5 (2001): 595–606.

White, Jerry. *London in the Nineteenth Century: 'A Human Awful Wonder of God'*. London: Vintage Books, 2013.

Williams, Rosalind H. *Dream Worlds: Mass Consumption in Late Nineteenth-Century France*. Berkeley: University of California Press, 1991.

Williams, Simon J. 'Modernity and the Emotions: Corporeal Reflections on the (Ir) Rational'. *Sociology* 32, no. 4 (1998): 747–69.

Wilson, Roger Burdett. *Go Great Western: A History of GWR Publicity*. Newton Abbot: David & Charles, 1970.

Wood, Karl E. *Health and Hazard: Spa Culture and the Social History of Medicine in the Nineteenth Century*. Newcastle: Cambridge Scholars Publishing, 2013.

Yeoman, Ian, Alastair Durie, Una McMahon-Beattie, and Adrian Palmer. 'Capturing the Essence of a Brand from Its History: The Case of Scottish Tourism Marketing'. *Journal of Brand Management* 13, no. 2 (2005): 134–47.
Zelizer, Viviana. *The Purchase of Intimacy*. Princeton: Princeton University Press, 2005.
Zuelow, Eric. *A History of Modern Tourism*. London: Palgrave Macmillan, 2015.

Books and articles

Ab-o'-th'-Yate [Benjamin Brierley]. *Drop't on at Blackpool*. Manchester: Abel Heywood & Son, 1881.
Ab-o'-th'-Yate [Benjamin Brierley]. *Adventures at Blackpool*. Manchester: John Heywood, 1872.
Anon. *The Scarborough Guide*. Scarborough: T. Coultas, 1815.
Anon. *Thorpe's Visitors' Hand-Book for Harrogate, and Strangers' Companion to All Objects of Interest in the District [electronic resource]: Together with a Popular Description of the Mineral waters, and a Reduced Ordnance Map of 20 Miles Round Harrogate, and Small Plan of the Town*. Harrogate: R. Ackrill, 1850.
Anon. 'The Long Vacation'. *British Medical Journal* 2, no. 710 (1874): 177.
Anon. *My Holiday: Where Shall I Spend It?* London: Sampson Low, Marston, Searle, and Rivington, 1878.
Anon. *Ward and Lock's Illustrated Guide to, and Popular History of Scarborough, Filey, Whitby, Btidlington Quay, and Their Vicinity*. London: Ward, Lock, and Co., 1880.
Anon. *The Dictionary of Watering Places, Seaside and Inland, at Home and Abroad*. London: L. Upcott Gill, 1881.
Anon. 'The English at the Seaside'. *William H. Harper's New Monthly Magazine* 63 (1881): 484–98.
Anon. 'The Wear and Tear of Modern Life'. *British Medical Journal* 1, no. 1318 (1886): 650–1.
Anon. *Adey's Annual Popular Guide to Scarborough*. London: Alexander & Shepheard, 1887.
Anon. *'Work and Leisure' Manuals to the Employment and Recreation for Women Workers. Part I: Where to Spend a Holiday*. London: Hatchards, 1887.
Anon. *The Advertiser's Guide to Publicity, etc.* Birmingham: Moody's Printing Co., 1887.
Anon. *Confessions of a Coquette, While Staying at Scarboro', Whitby, & Bridlington. By Azucena* Scarborough: Dennis, E. T. W, 1888.
Anon. *Visitors' Guide to Blackpool [Containing a Brief History of Blackpool, View of the Town, Street Plan and a Complete Guide to all Places of Interest or Pleasurable Resort in Blackpool and the Fylde]*. Blackpool: Blackpool Times, 1888.
Anon. 'Health in Relation to Work'. *The Lancet* 2, no. 3498 (1890): 570–1.
Anon. 'The Weekly Half-Holiday Bill'. *The Lancet* 1, no. 3473 (1890): 668–9.
Anon. *Illustrated Guide to Blackpool and District*. Blackpool: S. Wilson & Son, 1891.
Anon. 'Holiday Seekers'. *The Lancet* 2, no. 3651 (1893): 448.
Anon. *Holiday Notes in East Anglia. Being a Selection of Articles Which Have Appeared in the Public Press on the Holiday Resorts in Norfolk, Suffolk, and Essex; to Which Is Added Information Respecting Railway and Other Facilities for Intending Visitors*. Stratford: GER Company's Works, 1894.
Anon. 'The Increasing Nervousness of Our Time'. *British Medical Journal* 2, no. 1774 (1894): 1493.

Anon. 'The Sweets of Popularity'. *All the Year Round* 11, no. 271 (1894): 225–8.

Anon. 'British Balneological and Climatological Society'. *The Lancet* 1, no. 3799 (1896): 1723.

Anon. *Sunny Southsea: The Queen of Seaside Health Resorts*. Portsmouth: Holbrook & Son, 1896.

Anon. 'As Others See Us'. *The Journal of Balneology and Climatology* 1, no. 3 (1897): 291–4.

Anon. 'Editorials and Notes – Summer Holidays'. *The Journal of Balneology and Climatology* 1, no. 3 (1897): 249–51.

Anon. *Seaside Watering Places: A Description of Holiday Resorts on the Coasts of England and Wales, the Channel Islands, and the Isle of Man, Including the Gayest and the Quietest Places*. London: L. Upcott Gill, 1897.

Anon. *Blackpool the Unrivalled Sea Side Resort for Health & Pleasure*. Blackpool: Blackpool Corporation, 1898.

Anon. 'The Easter Holidays Extension'. *The Lancet* 1, no. 3996 (1900): 952.

Anon. 'The Proper Holiday'. *The Lancet* 2, no. 4067 (1901): 380.

Anon. *A New Pictorial and Descriptive Guide to Blackpool, Lytham, St. Anne's, Fleetwood, and The Fylde*. London: Ward, Lock & Co., 1902.

Anon. *Harrogate as a Health Resort: Published by the Corporation*. Harrogate: R. Ackrill, 1903.

Anon. *Blackpool Pictorial Guide*. Blackpool: J. P. Bamber, The Studio, 1904.

Anon. *Sally and Mary's Seaside Holiday Pictures*. London: R. Culley, 1907.

Anon. *At the Seaside*. London: Ward Lock & Co., 1908.

Anon. 'The Holiday Huddle'. *The Lancet* 2, no. 4534 (1910): 246.

Anon. 'The Claims of British Spas'. *The Lancet* 2, no. 4752 (1914): 804.

Anon. *The Beginnings of the Early Closing Movement*. London: Spottiswoode, Ballantyne & Co Ltd, Printers, 1917.

Alcock, George W. *Fifty Years of Railway Trade-Unionism*. London: Co-operative Printing Society Limited, 1922.

Amulree, William Warrender Mackenzie Baron. 'Industrial Holidays'. *Journal of the Royal Society of Arts* 87, no. 4486 (1938): 3–14.

Ash, Edwin. *Mind and Health: The Mental Factor and Suggestion in Treatment, with Special Reference to Neurasthenia and Other Common Nervous Disorders*. London: H. J. Glaisher, 1910.

Baker, Edward. *Sojourn in Scarborough 1850: The Diary of Edward Baker. With an Introduction by Patricia Fenteman*. Leeds: The Old Hall Press, 1984 [1850].

Beard, George. 'Neurasthenia, or Nervous Exhaustion'. *The Boston Medical and Surgical Journal* 80, no. 13 (1869): 217–21.

Beard, George Miller. *American Nervousness, Its Causes and Consequences: A Supplement to Nervous Exhaustion (Neurasthenia)*. New York: Putnam, 1881.

Beard, George M. *Practical Treatise on Nervous Exhaustion (Neurasthenia): Its Symptoms, Nature, Sequences, Treatment*. New York: E.B. Treat, 1899.

Benjamin, Walter. *The Arcades Project*. Translated by Howard Eiland. Cambridge: Harvard University Press, 1999 [1927–1940].

Besant, Walter. 'The Amusements of the People'. *Contemporary Review* 45 (1884): 342–53.

Blaker, Thomas F. I. *Burgess Hill as a Health Resort: Hints to Those in Search of a Healthy Country Home*. Brighton: John Beal and Co., 1883.

Burford, George and Clarence Granville Hey. *Practical Summary of Points to Be Noted in Cases of Neurasthenia in Women, Including Clinical History & Physical Finding*. London: J. Bale & Co., 1907.

Cheem, Aliph [Walter Yeldham]. *Lays of the Sea-Side: A Rhythmical Rendering of Sea-Side Stories and Incidents*. London: Edward Bumpus, 1887.

Clarke, Michell J. *Hysteria and Neurasthenia*. London: John Lane The Bodley Head, 1905.

Cleveland-Stevens, Edward. *English Railways, Their Development and Their Relation to the State*. London: George Routledge and Sons, 1915.

Clifford, John. *The Tunbridge Wells Guide*. Tunbridge Wells: J. Clifford, 1837.

Cole, Thomas H. (ed.). *Hastings and St. Leonards-on-Sea as a Health and Pleasure Resort. With Statistics & Local Information*. Hastings: F. J. Parsons, 1884.

Crichton-Browne, James. 'Abstract of an Address Delivered at the Opening of The Section of Psychology at the Meeting of the British Medical Association at Cambridge'. *The Lancet* 116, no. 2973 (1880): 292–5.

Davis, Hayton R. *Harrogate as a Health Resort*. Harrogate: R. Ackrill, 1889.

Dowse, Thomas S. *On Brain and Nerve Exhaustion. 'Neurasthenia.' Its Nature and Curative Treatment. A Paper Read before the Medical Society of London*. London: Bailliere, Tindall & Cox, 1880.

Doyle, Hy. M. 'Voyaging for Health: A Contribution to the Therapeutics of the Sea'. *The Lancet* 136, no. 3492 (1890): 226–8.

Drummond, David. 'Neurasthenia: Its Nature and Treatment'. *British Medical Journal* 2, no. 2375 (1906): 11–14.

Dutt, Aaron Chunder. *Health Notes for the Seaside: With Special Reference to Whitby and District*. Whitby: Horne & Son, 1895.

Farquharson, Robert. 'On Overwork'. *The Lancet* 107, no. 2731 (1876): 9–10.

Forman, H. S. *A Holiday Tour in and around Whitby*. Bradford: P. Lund & Co., 1896.

Gorky, Maxim. 'Boredom'. *Independent* 63, no. 8 (1907): 310–15.

Granville, Joseph Mortimer. '"Change" As a Mental Restorative'. *The Lancet* 115, no. 2963 (1880): 907–9.

Hartley, John. *Seets i' Blackpool, Fleetwood, Lytham, and Southport: As Seen bi Sammywell Grimes an' his wife Mally on their Hallidy Trip, wi' a Few Incidents an' Accidents 'at Occurred*. London: W. Nicholson & Sons, 1890.

Haviland, Alfred. *Scarborough as a Health Resort: It's Physical Geography, Geology, Climate, & Vital Statistics, With Health Guide Map, &c.* London: Hamilton, Adams and Co., 1883.

Holbeche, Arthur Oliver. 'The Claims of British Resorts, With Some Remarks on Malvern as a Health Resort'. *The Journal of Balneology and Climatology* 1, no. 4 (1897): 332–5.

Howells, William D. 'The Waters of Blackpool'. *The North American Review* 194, no. 673 (1911): 872–81.

Hyde, Samuel. 'Railways and British Health Resorts'. *The Journal of Balneology and Climatology* 1, no. 3 (1897): 176–80.

Johnson, George. 'Lectures on Some Nervous Disorders that result from Overwork and Mental Anxiety: Lecture I'. The Lancet 106, no. 2707 (1875): 85–7.

Johnson, George. 'Lectures on Some Nervous Disorders that result from Overwork and Mental Anxiety: Lecture III'. *The Lancet* 106, no. 2723 (1875): 651–4.

Keetley, C. B. 'On the Time and Place of Bathing and on the Hotel and Boarding-House Dietary at Health Resorts'. *The Journal of Balneology and Climatology* 1, no. 3 (1897): 222–4.

Kirkwood, S. 'Railway Tourist Literature'. *Railway Magazine* 6 (1900): 510–16.

Larking, Albert. *History of the Early Closing Association and How the Saturday Half-Holiday Was Won*. London: Unwin Brothers, Ltd, 1914.

Laycock, Arthur. *Fun on the Pleasure Beach wi' Sally an' Sam, Jimmy an' Jane Ann, towd by Sally hersel. A Funny Pennorth.* Blackpool: Laycock & Co., 1909.

Lewis, Francis David and Lowe. *The Advertiser's Guide to the Newspaper Press* [third edition]. London: Lewis and Lowe 1844.

MacFarlane, Aylmer. 'Some Remarks on Brighton as a Seaside Health Resort'. *The Journal of Balneology and Climatology* 1, no. 3 (1897): 232–5.

Mate, Charles H. and Charles Riddle. *Bournemouth: 1810–1910: The History of a Modern Health and Pleasure Resort.* Bournemouth: Messrs. W. Mate & Sons, Ltd, 1910.

Mitchell, S. Weir. *Rest in the Treatment of Nervous Disease.* New York: Putnam's Sons, 1875.

Mitchell, S. Weir. 'The Treatment by Rest, Seclusion, etc. in Relation to Psychotherapy'. *Journal of the American Medical Association* 50, no. 25 (1908): 2033–7.

Musgrove, Charles D. *Nervous Breakdowns and How to Avoid Them.* New York: Funk and Wagnalls Company, 1913.

Musgrove, Charles D. *Holidays and How to Use Them.* Bristol: J. W. Arrowsmith, 1914.

Nodal, John, H. and George Milner. *A Glossary of the Lancashire Dialect.* Manchester: Alexander Ireland & Co. 1875.

Oxenham, John. *The Very Short Memory of Mr Joseph Scorer and Other Seaside Experiences.* London: Simpkin, Marshall, & Co., 1903.

Poore, George V. 'The Climate of the Dwelling House'. *The Journal of Balneology and Climatology* 1, no. 4 (1897): 299–305.

Proust, Adrien A. and Gilbert Ballet. *The Treatment of Neurasthenia.* Translated by Peter Campbell Smith. London: Henry Kimpton, 1902.

Savage, George H. 'Overwork as a Cause of Insanity'. *The Lancet* 106, no. 2708 (1876): 127.

Savage, George H. 'Discussion on Neurasthenia and Its Treatment'. *British Medical Journal* (2), no. 1758 (1894): 522–5.

Sekon, G. A. 'The Cornwall Railway Then and Now'. *Railway Magazine* (1904).

Simmel, Georg. 'The Metropolis and Mental Life'. In Gary Bridge and Sophie Watson (eds), *The Blackwell City Reader*, pp. 103–10. Oxford: Wiley-Blackwell, 2002 [1903].

Shadwell, Arthur. *Industrial Efficiency: A Comparative Study of Industrial Life in England. Germany and America.* London: Longmans, Green and Co., 1906.

Short, One [Alfred Short]. *Adventures and Sketches at New Brighton by One Short.* Liverpool: A. Short, 1888.

Squire, WM. 'The Sulphur Springs of Great Britain and Their Therapeutic Action'. *The Lancet* 132, no. (1888): 201–3.

Thomson, Sinclair W. 'On Holiday-Making'. *The Lancet* 120, no. 3091 (1882): 912.

Tiebel, H. C. *Dr Tiebel's Official Handbook of Spas, Watering Places, and Health Resorts.* London: Charles Letts & Co., 1896.

Tuckwell, Gertrude. *Women's Work and Factory Legislation: The Amending Act of 1895, A Paper Read Before the Fabian Society.* London: Co-operative Printing Society Limited, 1895.

Veblen, Thorstein. *The Theory of the Leisure Class.* Oxford: Oxford University Press, 2007 [1899].

Webb, Sidney. *Women and the Factory Acts.* London: The Fabian Society, 1896.

Wilson, William S. *The Ocean as a Health-Resort: A Handbook of Practical Information as to Sea-Voyages for the Use of Tourists and Invalids.* London: J. & A. Churchill, 1880.

Wooton, Edwin. 'Where Will You Spend Your Holiday?' *London Society*, August 1898.

Yellowlees, David. 'An Address Delivered at the Opening of the Section of Psychology: The Causes and Prevention of Insanity'. *British Medical Journal* 2, no. 1283 (1885): 204–6.

Yeo, Isaac Burney. *Climate and Health Resorts.* London: Cassell & Company, 1890.

Index

Note: Figures are indicated by an italic '*f*' and notes are indicated by 'n' following the page numbers.

For the benefit of digital users, indexed terms that span two pages (e.g., 52–53) may, on occasion, appear on only one of those pages.